Optical and Wireless Communications

This edited volume brings together diverse perspectives on machine learning and AI applications in optical and wireless systems, offering a structured and comprehensive resource for researchers and professionals. It explores advancements driven by 5G, IoT, and the increasing demand for high-speed, reliable communication. Covering optical fiber systems, wireless networks, and AI-driven optimizations, the book provides insights into real-world applications impacting telecommunications, healthcare, and transportation.

The contributing authors discuss key topics such as signal processing techniques, optimization algorithms, and deep learning models applied to optical and wireless networks. The volume also highlights emerging challenges, security concerns, and future trends in AI-powered communication systems. This resource is essential for professionals in electrical and computer engineering, telecommunications, and computer science, helping them stay ahead in these rapidly evolving technologies.

Sushank Chaudhary (Senior Member IEEE, OPTICA, and SPIE) is an Associate Professor at Guangdong University of Petrochemical Technology. His research focuses on photonic radars, Visible Light Communication (VLC), Free Space Optics (FSO), and machine learning, with over 100 publications.

Amir Parnianifard (Senior Member IEEE) is an Assistant Professor at the University of Glasgow and UESTC. He specializes in computational intelligence and optimization, with extensive industry experience.

Abhishek Sharma (Senior Member IEEE, ECM OPTICA) is a faculty member at the National Institute of Technology (NIT) Hamirpur. His research includes photonic radars for autonomous vehicles and underwater navigation, VLC, FSO, and machine learning. He has over a decade of academic and research experience with 50+ publications.

Optical and Wireless Communications

Applications of Machine Learning and Artificial Intelligence

Edited by
Sushank Chaudhary, Amir Parnianifard, and Abhishek Sharma

CRC Press
Taylor & Francis Group
Boca Raton London New York

CRC Press is an imprint of the
Taylor & Francis Group, an **informa** business

Designed cover image: Data stream; Shutterstock

First edition published 2025
by CRC Press
2385 NW Executive Center Drive, Suite 320, Boca Raton FL 33431

and by CRC Press
4 Park Square, Milton Park, Abingdon, Oxon, OX14 4RN

CRC Press is an imprint of Taylor & Francis Group, LLC

ISBN: 9781032751245 (hbk)
ISBN: 9781032751252 (pbk)
ISBN: 9781003472506 (ebk)

DOI: 10.1201/9781003472506

Typeset in Minion
by Newgen Publishing UK

Contents

Editors' Bios

Sushank Chaudhary (Senior Member, IEEE, SM-OPTICA, M-SPIE) is an Associate Professor at the School of Computer, Guangdong University of Petrochemical Technology, Guangdong, China, where he has served since 2023. He received his Ph.D. from Universiti Utara Malaysia and has accumulated extensive postdoctoral experience, notably at the Fujian Institute of Research on the Structure and Matter, Chinese Academy of Sciences (2018–2020), and as a recipient of the prestigious C2F Fellowship at Chulalongkorn University, Bangkok, Thailand (2020–2023). Dr. Chaudhary has published over 100 research articles in leading international journals and conferences. His accolades include the Elite Talent Award from the Jinjiang Government in China (2019) and, in 2023, the Marie Skłodowska-Curie postdoctoral fellowship from the European Union, awarded with a Seal of Excellence for his research proposal on free-space optical technology. In addition to his research contributions, Dr. Chaudhary serves as an Associate Editor for *PLOS One*, *Frontiers in Physics*, and the *Journal of Sensors*. His research primarily focuses on autonomous vehicles, photonic radars, spatial division multiplexing, free-space optics, visible light communication systems, the metaverse, and digital twins.

Amir Parnianifard is an Assistant Professor with joint appointments at the University of Glasgow (UofG) and the University of Electronic Science and Technology of China (UESTC) since May 2022. He previously served as a Research Assistant at Chulalongkorn University, and in 2019, he earned his Ph.D. in Industrial Engineering from the University of Putra Malaysia. A Senior Member of IEEE and a certified Project Management Professional (PMP®), Dr. Parnianifard has authored numerous research articles and serves as a regular reviewer for various international journals. His research focuses on computational intelligence, surrogate-based modeling, uncertainty quantification, and optimization for engineering design. With over a decade of experience in the manufacturing industry, he is also recognized for his extensive teaching and student supervision.

Abhishek Sharma (Senior Member, IEEE; ECM-OPTIC) has been a faculty member at NIT Hamirpur since August 2024, specializing in photonic radars, visible light communication (VLC), and free-space optics (FSO). With over a decade of experience teaching undergraduate and postgraduate students at the university level, Dr. Sharma is deeply committed to advancing education and research in cutting-edge technologies.

Dr. Sharma's research primarily focuses on autonomous vehicles, photonic radars, spatial division multiplexing, free-space optics, visible light communication systems, the metaverse, and digital twins. He completed his Ph.D. in Electronics Technology from Guru Nanak Dev University (GNDU), Amritsar, from 2020 to 2024, focusing on the design of photonic radar systems for intelligent transportation.

His dedication to research excellence was recognized when he was awarded the prestigious Second Century Fund (C2F) Postdoctoral Fellowship in April 2024 to pursue advanced research at Chulalongkorn University, Thailand.

Dr. Sharma has published over 50 research articles in SCI- and Scopus-indexed international journals and conferences, amassing 1,100 citations and achieving an H-index of 20. His work continues to inspire advancements in photonic technologies and their applications.

Contributors

Geeta Arora earned her Ph.D. degree from IIT Roorkee, India, in 2011. She is working as a Professor in the Department of Mathematics at Lovely Professional University, Punjab, India.

Priya Chaudhary is working as a Research Scholar at Manipal University Jaipur. Her research interests are in the areas of optical communication, artificial intelligence, and the integration of AI into advanced communication systems.

Sushank Chaudhary (Senior Member, IEEE, SM-OPTICA, M-SPIE) is an Associate Professor at the School of Computer, Guangdong University of Petrochemical Technology, Guangdong, China, where he has served since 2023. He received his Ph.D. from Universiti Utara Malaysia and has accumulated extensive postdoctoral experience, notably at the Fujian Institute of Research on the Structure and Matter, Chinese Academy of Sciences (2018–2020), and as a recipient of the prestigious C2F Fellowship at Chulalongkorn University, Bangkok, Thailand (2020–2023).

Pinkey Chauhan is an Assistant Professor at Jaypee Institute of Information Technology based in Noida, Uttar Pradesh. Her research interest are in soft computing.

Divya C D is working as Associate Professor at the Department of Computer Science and Engineering, Vidyavardhaka College of Engineering, Mysore, Karnataka, India. Her research interests include optimization algorithms, machine learning, artificial intelligence, and their applications in wireless communication systems and vehicular networks. She is having 9 years of teaching experience.

Samarth V Gangadikar is a research scholar at the Department of Computer Science and Engineering, Vidyavardhaka College of Engineering, Mysore, Karnataka, India. His research interests include optimization algorithms, machine learning, artificial intelligence, and applications of artificial intelligence in wireless communication systems and vehicular networks.

Sonam Gour is working as a Research Scholar at Manipal University Jaipur. Her research interests are in the areas of optical communication, artificial intelligence, and the integration of AI into advanced communication systems.

Jaya Gupta completed her Ph.D. and working as a research scholar at the Department of Mathematics at Lovely Professional University, Punjab, India.

Pravesh Kumar is working as an Assistant Professor in the Department of Mathematics at Rajkiya Engineering College Bijnor (UP).

Nouman Latif is a research scholar at the Department of Electrical Engineering, Information Technology University, Lahore, Pakistan. His research focuses on optimization algorithms, AI applications, and integration of AI into modern communication systems.

Shubham Mishra is working as an Assistant Professor in the Department of Mathematics at Lovely Professional University, Punjab, India.

Srushti M. P. is a research scholar at the Department of Computer Science and Engineering, Vidyavardhaka College of Engineering, Mysore, Karnataka, India. Her research interests include optimization algorithms, machine learning, artificial intelligence, and applications of artificial intelligence in wireless communication systems and vehicular networks.

Hum Nath Parajuli (Member, IEEE) received his M.Sc. degrees from Osaka University, Japan, and Scuola Superiore Sant'Anna, Italy, through the EU-funded Erasmus Mundus Scholarship Program, and his Ph.D. degree from the Budapest University of Technology and Economics, Hungary, under the EU-funded Marie Curie ITN program, specializing in photonics engineering.

Amir Parnianifard is an Assistant Professor with joint appointments at the University of Glasgow (UofG) and the University of Electronic Science and Technology of China (UESTC) since May 2022. He previously served as a Research Assistant at Chulalongkorn University, and in 2019, he earned his Ph.D. in Industrial Engineering from the University of Putra Malaysia.

Amit Rathi is working as a Professor at Manipal University Jaipur. His research interests are in the areas of optical communication, artificial intelligence, and the integration of AI into advanced communication systems.

Saanvi Ravi is a research scholar at the Department of Computer Science and Engineering, Vidyavardhaka College of Engineering, Mysore, Karnataka, India. His research interests include optimization algorithms, machine learning, artificial intelligence, and applications of artificial intelligence in wireless communication systems and vehicular networks.

Demostenes Zegarra Rodriguez is a Professor, Department of Computer Science, Federal University of Lavras, Brazil. His research interests include artificial intelligence, deep learning, and neural networks, with a focus on their applications in various communication systems.

Muhammad Saadi is an Associate Professor and Head of the Electrical Engineering Department, Faculty of Engineering, University of Central Punjab. He received his Ph.D. from Chulalongkorn University, Thailand, and has held roles as a research assistant at NECTEC, lecturer at KMUTNB, and technical content writer at Aimagin.

Samjith T S is a research scholar at the Department of Computer Science and Engineering, Vidyavardhaka College of Engineering, Mysore, Karnataka, India. His research interests include optimization algorithms, machine learning, artificial intelligence, and applications of artificial intelligence in wireless communication systems and vehicular networks.

Abhishek Sharma has been a faculty member at NIT Hamirpur since August 2024, specializing in photonic radars, visible light communication (VLC), and free-space optics (FSO). He completed his Ph.D. in Electronics Technology from Guru Nanak Dev University (GNDU), Amritsar, from 2020 to 2024, focusing on the design of photonic radar for intelligent transportation systems.

Shivani Sharma is currently working as an Assistant Professor in the Computer Science and Engineering Department at Thapar Institute of Engineering & Technology (Deemed University). She has done her Ph.D. in Computer Science and Engineering Department from Indian Institute of Technology Roorkee, Master of Engineering from Indian Institute of information Technology and Management Gwalior (IIITM), India, and B.Tech. from UPTU India. Her main research interests are data masking for privacy preservation, data mining, and machine learning.

Pushkar V is a research scholar at the Department of Computer Science and Engineering, Vidyavardhaka College of Engineering, Mysore, Karnataka, India. His research interests include optimization algorithms, machine learning, artificial intelligence, and applications of artificial intelligence in wireless communication systems and vehicular networks.

Megha Varshney is working as a research scholar in the Department of Mathematics at Rajkiya Engineering College Bijnor (UP) and has registered for Ph.D. at Dr. A.P.J. Abdul Kalam Technical University, Lucknow.

Avani Vyas is working with the Department of Information Technology at National Institute of Technology, Jalandhar, India, and has done her Ph.D. from IIT Ropar. Her area of expertise are smart healthcare, E-health, networking and communication, and data analysis.

CHAPTER 1

Introduction to Advances in Optical and Wireless Communication

Role of Machine Learning and AI

Abhishek Sharma, Sushank Chaudhary, and Amir Parnianifard

1.1 INTRODUCTION

1.1.1 Overview of Optical and Wireless Communication

Optical and wireless communications are the two pillars of modern telecommunication networks, enabling the rapid and reliable transmission of data across the globe. Optical communication primarily relies on light waves to transmit information through optical fibers, providing high-speed, long-distance communication with minimal loss and interference. Since its inception in the 1970s, optical fiber technology has revolutionized the way data is transmitted, enabling the exponential growth of the internet, video streaming, and other data-intensive applications [1–4].

Wireless communication, on the other hand, relies on radio frequency (RF) waves to transmit data over the air [5,6]. This technology has evolved

DOI: 10.1201/9781003472506-1

significantly since the advent of wireless telegraphy in the late 19th century, leading to the development of mobile telephony, Wi-Fi, satellite communication, and the latest 5G networks. Wireless communication offers the flexibility of mobility and has become indispensable in today's connected world, enabling everything from smartphone connectivity to the internet of things (IoT) [7–11].

The combination of the optical with wireless has been vital in satisfying the continuous increment in the demand for bandwidth and connectivity. Optical is used in transmitting data at higher speed and to connect a location with other locations, which are required for the backbone of communication, while wireless technologies are used to deliver this connectivity to the customers [12–14]. As a whole, these technologies are fundamental to today's communication platforms for a multiplicity of applications ranging from casual internet browsing to essential mission operations.

1.1.2 Emerging Trends and Challenges

The growth of the new digital era together with the immense development of new devices enriched the field of both optical and wireless communication. The trends such as 5G networks, cloud services and particularly IoT appear to be requiring further development of these areas.

In optical communication, advancements in wavelength-division multiplexing (WDM) and the development of coherent optical technologies have enabled higher data rates and more efficient use of the optical spectrum [15–20]. These technologies are crucial for supporting the massive data traffic generated by modern applications. Additionally, innovations in free-space optical (FSO) communication are opening new possibilities for high-speed, line-of-sight communication in environments where traditional fiber optics are impractical [21–23].

Communication without wire is not an exception to this wave; modern techniques like massive MIMO (multiple input multiple output) [24–31], millimeter-wave communication, beamforming are some of the techniques involved in the 5G network. These technologies seek to build the capability, scope, and effectiveness of wireless networks based on the increasing need of higher data rate mobility. In addition, AI as well as machine learning (ML) holds sweeping potential for the application in enhancing wireless networks in terms of aspects such as resource management, interference, and security.

1.2 FUNDAMENTALS OF OPTICAL AND WIRELESS COMMUNICATION

1.2.1 Optical Communication

Optical communication refers to the transmission of data through light signals, typically via optical fibers. This kind of communication has become the fundamentals of modern telecommunication networks since it carries bulk volume of data over long distance with low loss.

1.2.1.1 Basic Principles

The core principle of optical communication is the modulation of light to carry information. Light, often generated by lasers or light-emitting diodes (LEDs), is modulated to encode data, which is then transmitted through optical fibers [2,4,32]. These fibers are made of glass or plastic and have a core surrounded by cladding, which ensures that the light remains trapped within the core, allowing it to travel long distances with minimal attenuation [33].

1.2.1.2 Key Technologies

Several technologies have been developed to enhance the performance of optical communication systems:

- **Wavelength-Division Multiplexing (WDM):** Wavelength-division multiplexing (WDM) is a technique that enhances the data transmission capability of optical fibers by allowing multiple light wavelengths to be sent simultaneously through a single fiber. Each distinct wavelength conveys its own data stream, thereby significantly increasing the overall capacity of the fiber [15–18,34,35].
- **Passive Optical Networks (PON):** Passive optical networks (PONs) are employed to provide broadband network access to users. Utilizing a point-to-multipoint configuration, they allow a single optical fiber to connect and serve multiple endpoints, including homes or businesses [36–38].
- **Coherent Optical Transmission:** This method involves adjusting both the amplitude and phase of light waves to enhance the data

transmission capacity. Coherent systems are well-suited for backbone networks due to their ability to support higher data rates and extend transmission distances [32,39].

1.2.1.3 Applications in Modern Networks

Optical communication plays a vital role in various aspects of modern networks:

- **Backbone Networks:** Optical fibers form the backbone of global telecommunications infrastructure, connecting data centers, internet exchanges, and other critical network components.
- **Data Centers:** High-speed optical links within and between data centers ensure fast and reliable data transfer, supporting cloud computing, big data analytics, and other data-intensive applications.
- **5G Backhaul:** As 5G networks are deployed, optical communication is essential for backhauling data from cell sites to the core network, supporting the high data rates and low latency required by 5G applications.

1.2.2 Wireless Communication

Wireless communication enables the transmission of data over the air using radio waves, microwaves, or millimeter waves. It is the foundation of mobile networks, Wi-Fi, satellite communication, and many other wireless technologies that we rely on daily.

1.2.2.1 Basic Principles

Wireless communication involves the transmission of electromagnetic waves between a transmitter and a receiver. Data is encoded onto these waves through various modulation techniques, such as amplitude modulation (AM), frequency modulation (FM), and phase modulation (PM). The wireless medium, being more susceptible to interference and noise compared to optical fibers, requires advanced techniques to maintain signal integrity.

1.2.2.2 Key Technologies

Several critical technologies have been developed to enhance the performance and reliability of wireless communication systems:

- **Multiple Input Multiple Output (MIMO):** Multiple input multiple output (MIMO) technology improves wireless communication performance by employing several antennas at both the transmitting and receiving ends. This setup enables the simultaneous transmission of multiple data streams, thereby enhancing the capacity and reliability of wireless networks [21,31,40–42].
- **Beamforming:** Beamforming is a technique that focuses the transmission or reception of signals toward specific directions, rather than dispersing them uniformly in all directions. This approach improves signal quality and minimizes interference, particularly in areas with a high concentration of users [43–45].
- **Millimeter-Wave Communication:** Millimeter-wave communication operates within the frequency range of 30–300 GHz, providing significantly higher data rates and bandwidth compared to conventional microwave communication. As a crucial component of 5G networks, it delivers the high-speed connectivity necessary for advanced applications such as augmented reality, virtual reality, and autonomous vehicles [18,19].

1.2.2.3 Challenges in Spectrum Management and Interference

Spectrum management is a critical issue in wireless communication. The RF spectrum is a finite resource, and the increasing demand for wireless services has led to congestion in many frequency bands. Efficient spectrum management involves allocating frequencies to different services and users in a way that minimizes interference and maximizes spectrum utilization.

Interference, both from other wireless devices and environmental factors, is another significant challenge. Techniques such as adaptive modulation, error correction, and interference cancellation are used to mitigate the effects of interference and ensure reliable communication.

1.2.2.4 Role in Modern Networks

Wireless communication is ubiquitous in modern networks:

- **Mobile Networks:** Wireless communication is the backbone of mobile networks, enabling voice, data, and multimedia services to billions of users worldwide.

- **Wi-Fi:** Wi-Fi networks provide wireless internet access in homes, offices, and public spaces, connecting a wide range of devices from smartphones to smart home appliances.
- **Satellite Communication:** Satellite communication extends wireless connectivity to remote and underserved areas, providing critical services like weather forecasting, GPS, and global broadcasting.

1.3 ROLE OF MACHINE LEARNING AND AI IN OPTICAL COMMUNICATION

1.3.1 Introduction to Machine Learning and AI

ML and artificial intelligence (AI) have emerged as powerful tools in modern communication systems, offering the potential to significantly enhance performance, optimize operations, and address complex challenges that are difficult to solve with traditional approaches [46–49]. In the context of optical communication, ML and AI techniques are being increasingly adopted to improve various aspects of the system, from signal processing and network optimization to fault detection and management.

ML involves the use of algorithms that can learn from and make predictions or decisions based on data. AI, a broader concept, encompasses ML and other techniques that enable machines to perform tasks that would typically require human intelligence, such as pattern recognition, decision-making, and problem-solving. The integration of ML and AI into optical communication systems is driven by the need for greater efficiency, adaptability, and resilience in the face of growing network complexity and data traffic demands.

1.3.2 Applications of Machine Learning in Optical Communication

1.3.2.1 Signal Processing

One of the primary areas where ML has made a significant impact in optical communication is in signal processing [48,50–52]. Optical signals are often subject to various impairments, such as noise, dispersion, and nonlinearities, which can degrade signal quality and limit system performance. Traditional signal processing techniques may struggle to compensate for these impairments, especially as data rates increase and system complexity grows.

ML offers a more adaptive and data-driven approach to signal processing. Techniques such as deep learning, which involves training neural networks on large datasets, can be used to model and compensate for complex signal distortions. For example, ML algorithms can be trained to predict and correct nonlinear distortions in optical fibers, improving signal quality and enabling higher data rates.

For instance, a specific case study involves applying deep learning techniques to enhance coherent optical communication systems. By using AI to optimize the signal processing pipeline, researchers were able to significantly reduce signal distortion and increase data throughput in long-haul fiber optic networks. Another example could be the application of AI in managing WDM systems, where ML models predict optimal wavelength allocation to minimize crosstalk and interference, thereby improving overall network efficiency.

1.3.2.2 Network Optimization

Network optimization is another critical area where ML is being applied in optical communication. Optical networks are complex and dynamic, with traffic patterns that can vary significantly over time. Efficiently managing resources, such as bandwidth and wavelength assignments, is crucial to maximizing network performance and minimizing costs [53,54].

ML algorithms can be employed to predict traffic patterns, optimize routing, and dynamically allocate resources in optical networks. For example, ML-based predictive models can forecast network congestion and proactively adjust routing to avoid bottlenecks. This dynamic optimization helps to ensure that network resources are used efficiently, reducing latency and improving overall service quality [55,56].

Moreover, AI-driven network management systems can automate routine tasks, such as fault detection and recovery, reducing the need for manual intervention and minimizing the risk of human error. These systems can continuously monitor network performance, detect anomalies, and take corrective actions in real-time, ensuring that the network operates at peak efficiency.

1.3.2.3 Fault Detection and Management

Fault detection and management are crucial aspects of maintaining the reliability and availability of optical communication networks. Failures in optical fibers, amplifiers, or other components can lead to significant disruptions in service, making it essential to detect and address issues as quickly as possible [11].

ML and AI techniques are increasingly being used to enhance fault detection and management in optical networks. By analyzing vast amounts of network data, ML algorithms can identify patterns that indicate potential faults or degradation in performance. For instance, AI systems can detect subtle changes in signal quality that may precede a component failure, allowing for predictive maintenance and reducing downtime [57].

In addition to fault detection, ML and AI can also assist in fault localization and recovery. Once a fault is detected, AI algorithms can quickly determine its location within the network and suggest the most effective recovery actions. This automation not only speeds up the resolution process but also reduces the operational costs associated with manual troubleshooting [58].

1.4 ROLE OF MACHINE LEARNING AND AI IN WIRELESS COMMUNICATION

1.4.1 Enhancing Wireless Communication with Machine Learning

Wireless communication systems have evolved significantly over the past few decades, driven by the increasing demand for higher data rates, improved coverage, and better quality of service (QoS). As these systems become more complex, traditional methods of managing and optimizing wireless networks are reaching their limits. ML and AI have emerged as powerful tools to address these challenges, offering new ways to enhance the performance and efficiency of wireless communication systems.

1.4.1.1 Spectrum Management

Spectrum is a critical resource in wireless communication, and its efficient management is essential for maximizing network performance. As the number of connected devices continues to grow, the competition for spectrum resources intensifies, leading to congestion and interference. Traditional static spectrum allocation methods are increasingly inadequate in this dynamic environment [59].

ML can revolutionize spectrum management by enabling dynamic and intelligent spectrum access. Cognitive radio networks (CRNs) are a prime example of this application. In a CRN, wireless devices are equipped with cognitive capabilities that allow them to sense the spectrum environment, identify unused spectrum (often referred to as "white spaces"), and adapt their transmission parameters accordingly. ML algorithms can enhance these cognitive capabilities by enabling more accurate spectrum sensing, prediction of spectrum availability, and decision-making processes for spectrum access [60,61].

For instance, reinforcement learning, a type of ML, can be used to develop intelligent spectrum access strategies. By continuously interacting with the environment and learning from the outcomes of its actions, a reinforcement learning algorithm can optimize spectrum usage, minimizing interference and maximizing throughput.

1.4.1.2 Channel Estimation and Modeling

Accurate channel estimation is crucial for reliable wireless communication, particularly in environments with high mobility or multipath propagation. Traditional channel estimation techniques, such as pilot-based methods, can be limited in their ability to cope with rapidly changing channel conditions or complex propagation environments [62].

ML offers a data-driven approach to channel estimation, where models can be trained to predict channel characteristics based on historical data. Deep learning, in particular, has shown promise in this area, with neural networks being used to model complex channel behaviors and improve estimation accuracy. These models can adapt to changing conditions, providing more robust and accurate channel estimates, which in turn enhances the performance of techniques like MIMO and beamforming [63–65].

For example, conventional channel estimation might rely on fixed pilot symbols and predefined algorithms, which may not adapt well to rapidly changing environments like urban canyons or high-speed vehicular communication. In contrast, a deep learning model trained on diverse channel conditions can dynamically adjust its parameters in real time, improving estimation accuracy by up to 30% compared to traditional methods. This improved estimation directly translates to more efficient use of the spectrum and better overall network performance.

1.4.1.3 Network Security

As wireless networks become more pervasive and critical to everyday life, ensuring their security is of paramount importance. Traditional security mechanisms, such as encryption and authentication, are essential but may not be sufficient to protect against the increasingly sophisticated cyber threats targeting wireless communication systems [66].

ML and AI can significantly enhance wireless network security by enabling more advanced threat detection and response capabilities. Intrusion detection systems (IDS), which monitor network traffic for suspicious activities, can be greatly improved with ML algorithms. These algorithms can analyze vast amounts of traffic data to detect patterns indicative of potential attacks, such as unauthorized access attempts, distributed denial-of-service (DDoS) attacks, or data exfiltration [67].

For example, anomaly detection techniques, a subset of ML, can be used to identify deviations from normal traffic patterns that may signal a security breach. Once an anomaly is detected, AI-driven systems can automatically initiate countermeasures, such as isolating the affected network segment or blocking malicious traffic [68–70].

Additionally, ML can be used to enhance encryption techniques by optimizing key management processes and identifying vulnerabilities in cryptographic algorithms. By leveraging AI, wireless networks can become more resilient to emerging threats, ensuring the confidentiality, integrity, and availability of data.

1.4.2 Integration of AI in Advanced Wireless Technologies

The integration of AI in wireless communication is not limited to spectrum management, channel estimation, and security. It extends to various advanced technologies that are shaping the future of wireless networks, particularly in the context of 5G and beyond.

1.4.2.1 Massive MIMO and Beamforming

Massive MIMO is considered as a crucial technology in today's wireless networks and a key element of 5G networks. The mass antenna configuration at the base station enables it to support a multitude of users while at the same time increasing capacity and spectral efficiency. Nonetheless, the realization of these benefits is undermined by two key challenges that arise with the massive MIMO systems; these are channel estimation, signal processing, and resource allocation [71–73].

These elements of massive MIMO are found to have tremendous promise in being improved by the application of ML. For instance, deep learning algorithms can be used to improve the estimate of channel, thus minimizing the use of pilots and increasing the general performance of a system. Also, ML can be applied for enhancement of beamforming where the signals are directed to or received from specific directions with a view of enhancing the quality of the signals as well as reducing the effects of interference. Using real-time inputs, the algorithms can then continuously optimize beamforming plan to accommodate the changing circumstances and maintain the most effective one at all times.

1.4.2.2 Millimeter-Wave Communication

Millimeter-wave communication, which operates at frequencies between 30 and 300 GHz, is another critical technology in 5G and beyond. It offers much higher data rates and bandwidth compared to traditional microwave communication but also faces challenges such as high path loss, susceptibility to blockages, and limited range [74–76].

ML can help address these challenges by enabling more efficient resource management and adaptive transmission strategies. For example, AI algorithms can predict the likelihood of signal blockages (such as from obstacles like buildings or trees) and adjust the transmission parameters accordingly, improving the reliability and performance of millimeter-wave communication. Additionally, ML can be used to optimize the handover process in millimeter-wave networks, ensuring seamless connectivity as users move between different cells.

1.5 INTEGRATION OF OPTICAL AND WIRELESS COMMUNICATION USING AI

1.5.1 Convergence of Optical and Wireless Networks

Networking of optical and wireless systems is a new trend in the telecommunication market that developed due to the need for high-speed, reliable connection of users all over the world. This integration is a two-pronged tactic that builds on the features of optical and wireless technologies to provide greater effectiveness in a system of communications.

Optical communication enables high bandwidth and low latency requirement that are suitable for communication network backbones and long-distance high-capacity data transmission [77]. Wireless communication on the other hand makes it possible to have mobility and flexibility for last-mile connectivity to allow end-users easy access to the internet. When these two technologies are implemented in a network, the latter can feature both the high speed of the optical communication as well as the ease and ubiquity of the wireless communication [78].

AI's role in the convergence of optical and wireless networks goes beyond basic traffic management [79,80]. It involves leveraging predictive analytics to forecast network demand and proactively allocate resources, ensuring that both the optical and wireless segments operate at peak efficiency. For instance, AI can predict high-traffic periods in a mobile network and preemptively allocate additional optical bandwidth to handle the increased load, thereby preventing congestion before it occurs.

1.5.1.1 Benefits of Integrated Networks

The integration of optical and wireless networks offers several benefits:

- **Enhanced Network Performance:** The combined use of optical and wireless technologies can significantly enhance network performance by optimizing data transmission across different segments of the network. Optical networks can handle the bulk of data traffic, while wireless networks provide flexible, high-speed access to end-users.
- **Improved Scalability:** Integrated networks are more scalable, allowing for easier expansion as demand grows. The use of AI enables dynamic resource allocation, ensuring that the network can adapt to varying traffic loads and user demands.
- **Cost Efficiency:** By leveraging the strengths of both technologies, integrated networks can reduce operational costs. For example, wireless networks can offload data to optical networks during periods of high demand, reducing the need for expensive wireless infrastructure upgrades.

1.5.2 AI-Driven Integration Techniques

AI plays a critical role in the successful integration of optical and wireless networks. It enables the development of intelligent systems that can manage and optimize the complex interactions between these two types of communication technologies.

1.5.2.1 Traffic Management and Optimization

Managing data flow between optical and wireless segments is a significant challenge in integrated networks. AI-powered traffic management systems can overcome this by continuously monitoring network conditions and dynamically routing data to the most efficient segment, thereby enhancing performance and reducing congestion.

For instance, AI algorithms can predict traffic trends and allocate resources accordingly across the network. In times of high traffic, the system might prioritize latency-sensitive data by routing it through the optical network, while less critical data is managed by the wireless segment. This adaptive approach ensures that both network segments maintain peak efficiency.

The predictive abilities of AI, such as using ML models to foresee network congestion before it occurs, are crucial in these integrated systems. For example, during a large-scale live event, AI can analyze historical and real-time data to predict spikes in usage, allowing the system to adjust resource allocation between the optical and wireless segments to maintain optimal network performance.

1.5.2.2 Resource Allocation

In integrated networks, efficient resource allocation is essential to maintaining high levels of performance and reliability. AI can be used to develop advanced resource allocation algorithms that consider a wide range of factors, including network load, user demand, and environmental conditions.

For example, AI can optimize the allocation of bandwidth between the optical and wireless segments, ensuring that each segment has sufficient resources to meet user demands. It can also manage the handover process in hybrid networks, ensuring that users experience seamless connectivity as they move between different segments.

1.5.2.3 Network Slicing

Network slicing is a technique that enables the creation of multiple virtual networks within a single physical infrastructure, each tailored to meet the specific requirements of different applications or services. This is particularly useful in integrated networks, where different types of traffic (e.g., IoT data, video streaming, voice calls) have different performance requirements.

AI can enhance network slicing by dynamically adjusting the configuration of each slice based on real-time data. For example, during a live sports event, the system might allocate more resources to the slice handling video streaming, ensuring high-quality video delivery to viewers. After the event, resources can be reallocated to other slices, such as those supporting IoT devices or mobile users.

1.5.3 Use Cases and Applications

The integration of optical and wireless networks, powered by AI, is opening up new possibilities across various industries and applications. Here are a few examples:

1.5.3.1 Smart Cities

Smart cities rely on a vast array of sensors and devices that generate large amounts of data. Integrated optical and wireless networks can provide the necessary infrastructure to support these devices, with optical networks handling high-capacity data transmission and wireless networks offering flexible, low-latency connectivity. AI can further optimize these networks, ensuring efficient data collection, processing, and distribution [81,82].

For example, AI can manage traffic data from smart city sensors, optimizing traffic light patterns to reduce congestion and improve safety. It can also ensure that critical data, such as emergency response information, is prioritized and delivered quickly and reliably.

1.5.3.2 Enhanced Mobile Broadband (eMBB)

Enhanced mobile broadband is one of the key use cases of 5G, providing high-speed internet access to users, even in dense urban environments. Integrated networks can support eMBB by combining the high capacity of optical networks with the mobility and flexibility of wireless networks [83,84].

AI can optimize the delivery of eMBB services by managing the handover process between the optical and wireless segments, ensuring seamless connectivity and high data rates. It can also predict user demand and allocate resources accordingly, preventing congestion and maintaining high-quality service.

1.5.3.3 Ultra-Reliable Low-Latency Communication (URLLC)

Ultra-reliable low-latency communication (URLLC) is a crucial 5G application, especially for scenarios like self-driving cars, telemedicine procedures, and industrial automation, where high reliability and very low latency are vital. Integrated networks are equipped to meet these demands by combining the low-latency data transmission of optical networks with the real-time communication capabilities of wireless networks [85,86].

AI can enhance URLLC by predicting potential network issues and proactively adjusting network configurations to maintain reliability and low latency. For example, in an autonomous vehicle network, AI can ensure that data from sensors and control systems is transmitted with minimal delay, ensuring safe and efficient operation.

1.5.4 Challenges and Future Directions

While the integration of optical and wireless networks offers many benefits, it also presents several challenges. One of the primary challenges is ensuring seamless interoperability between the two types of networks, which often use different technologies and protocols. AI can help address this challenge by providing intelligent management and optimization solutions, but the development of these solutions requires significant research and innovation.

Another challenge is ensuring the security of integrated networks, particularly as the attack surface increases with the convergence of multiple technologies. AI-driven security systems can help mitigate these risks, but they must be carefully designed and continuously updated to stay ahead of emerging threats.

1.6 STATE OF THE ART IN OPTICAL AND WIRELESS COMMUNICATION

The landscape of optical and wireless communication is continuously evolving, with rapid advancements driven by the need for higher data rates, increased capacity, and more reliable connectivity. This section provides an overview of the latest developments in these fields, highlighting the state of the art in both optical and wireless communication technologies.

1.6.1 Recent Advances in Optical Communication

Optical communication has seen significant innovations in recent years, with new technologies and techniques pushing the boundaries of data transmission capacity, efficiency, and reach. These advancements are crucial for meeting the growing demands of modern communication networks, particularly in the context of data centers, backbone networks, and 5G infrastructure.

1.6.1.1 Coherent Optical Communication

Coherent optical communication is one of the most significant developments in the field. Unlike traditional intensity-modulation direct-detection (IM-DD) systems, coherent communication systems use both the amplitude and phase of the light wave to encode data, significantly increasing the data capacity of optical links. This technology has become the standard for long-haul optical transmission and is now being adopted in shorter-reach applications, such as data center interconnects [87–91].

Coherent systems also benefit from advanced digital signal processing (DSP) techniques, which help to mitigate impairments like chromatic dispersion and polarization mode dispersion. The combination of coherent detection and sophisticated DSP has enabled transmission rates of 400 Gbps and beyond on a single wavelength, with research now focusing on pushing these rates even higher [92–94].

1.6.1.2 Space-Division Multiplexing (SDM)

As the demand for data continues to grow, traditional WDM systems are approaching their capacity limits. Space-division multiplexing (SDM) is an emerging technology that offers a solution by increasing the number of spatial channels available for data transmission [95–97].

SDM can be implemented using multi-core fibers (MCFs) or few-mode fibers (FMFs), which allow multiple data streams to be transmitted simultaneously through separate spatial channels. This approach significantly enhances the overall capacity of optical networks without requiring additional bandwidth. SDM is still in the research phase, but it holds great promise for future optical communication systems, particularly in scenarios where extreme data capacity is required.

1.6.1.3 Optical Access Networks and PON Technologies

Passive optical networks (PONs) are critical for delivering high-speed broadband access to end-users. Recent advancements in PON technology have focused on increasing the data rates and reach of these networks. Next-generation PONs (NG-PON2) are now being deployed, offering symmetrical data rates of up to 10 Gbps and support for multiple wavelengths, allowing for greater flexibility and capacity [91,98,99].

Research is also ongoing into new modulation formats and coding schemes to further enhance the performance of PONs, as well as the development of software-defined networking (SDN) and network function virtualization (NFV) solutions for more dynamic and programmable optical access networks.

1.6.1.4 Free-Space Optical (FSO) Communication

FSO communication is gaining renewed interest as a complementary technology to traditional fiber optic networks. FSO uses light to transmit data through the air, offering high data rates and low latency over short distances. It is particularly useful in situations where laying optical fibers is impractical or cost-prohibitive, such as in urban areas, remote locations, or disaster recovery scenarios.

Recent advancements in FSO technology include the development of adaptive optics to mitigate atmospheric turbulence, as well as hybrid FSO/RF systems that combine the strengths of optical and radio frequency communication to improve reliability and performance [2,15,75, 100–103].

In the last five years, FSO communication has seen significant improvements in both hardware and software, enabling more reliable and higher-capacity links [13, 14,87,104–111]. Adaptive optics systems, initially developed for astronomy, are now being repurposed for FSO to correct for distortions caused by atmospheric conditions, leading to more stable communication links even in adverse weather [112].

1.6.2 Recent Advances in Wireless Communication

Wireless communication technologies are evolving rapidly, driven by the deployment of 5G networks and the ongoing research into 6G. These advancements are aimed at addressing the challenges of higher data rates, lower latency, and more reliable connectivity in increasingly dense and heterogeneous network environments.

1.6.2.1 5G and Beyond

The rollout of 5G networks is revolutionizing wireless communication by delivering significantly faster data rates, reduced latency, and greater capacity compared to earlier generations. Several key technologies underpin the functionality of 5G, including massive MIMO, millimeter-wave communication, and network slicing.

- **Massive MIMO:** This technology leverages a large array of antennas at base stations to simultaneously serve multiple users, thereby increasing spectral efficiency and network capacity. ML advancements are further enhancing massive MIMO systems by optimizing processes such as beamforming and channel estimation [24].

- **Millimeter-Wave Communication:** Operating within the 30–300 GHz frequency range, millimeter-wave communication offers the high bandwidth essential for 5G services. Recent innovations have focused on addressing challenges like signal attenuation and blockage, with solutions such as advanced beamforming techniques and AI-driven adaptive transmission strategies [113,114].
- **Network Slicing:** This technology enables the creation of multiple virtual networks on a single physical infrastructure, each customized for specific applications (e.g., enhanced mobile broadband, IoT, URLLC). This adaptability is vital for accommodating the wide variety of use cases within 5G networks.

1.6.2.2 AI-Enhanced Wireless Networks

The incorporation of AI and ML into wireless networks is significantly changing the landscape of network management and optimization. AI-driven strategies are improving several vital aspects of wireless communication, including:

- **Adaptive Spectrum Management:** AI algorithms are enhancing spectrum efficiency by dynamically adjusting frequency allocations based on real-time demands and interference patterns.
- **Autonomous Network Optimization (SON):** AI is central to the development of SONs, which automatically optimize network parameters like power settings, frequency usage, and handover protocols. This automation boosts network performance and lowers operational expenses.
- **Enhanced Security Measures:** AI is being utilized to fortify wireless network security, enabling the real-time identification and mitigation of threats, thereby protecting the integrity and confidentiality of data transmission.

1.6.2.3 Integration with Edge Computing

Edge computing is becoming increasingly important in wireless communication, particularly with the rollout of 5G and the development of IoT. By processing data closer to the source, edge computing reduces latency and bandwidth usage, improving the performance of latency-sensitive applications such as autonomous vehicles, remote surgery, and real-time analytics.

Recent advancements in this area include the development of AI-driven edge computing platforms that can dynamically allocate resources, optimize data processing, and ensure seamless integration with core networks.

1.6.3 Case Studies

The state-of-the-art technologies in optical and wireless communication are being applied in a variety of real-world scenarios, demonstrating their potential to transform industries and improve connectivity.

1.6.3.1 Smart Transportation Systems

Smart transportation systems are leveraging both optical and wireless communication technologies to enable connected vehicles, intelligent traffic management, and real-time data sharing. For example, advanced optical networks provide the backbone for data transmission, while 5G wireless networks enable vehicle-to-everything (V2X) communication, improving safety and efficiency on the roads [115–118].

1.6.3.2 Industrial Automation and Industry 4.0

In the context of Industry 4.0, factories and industrial environments are becoming increasingly connected through IoT devices and wireless communication networks. Optical communication supports the high-speed transfer of large datasets generated by industrial processes, while wireless networks enable real-time monitoring and control. AI plays a crucial role in optimizing these integrated systems, ensuring that data is processed and acted upon quickly and efficiently.

1.6.3.3 Telemedicine and Remote Healthcare

Telemedicine and remote healthcare services are expanding rapidly, driven by advancements in communication technologies. High-speed optical networks support the transmission of medical imaging and other large datasets, while wireless networks enable real-time communication between patients and healthcare providers. AI-driven solutions enhance these services by enabling predictive diagnostics, remote monitoring, and personalized treatment plans [12,89,119–122].

1.7 CHALLENGES AND FUTURE DIRECTIONS

The rapid advancements in optical and wireless communication technologies, particularly with the integration of ML and AI, have transformed the landscape of global communication networks. However, these innovations come with a set of challenges that need to be addressed to fully realize their potential. Additionally, there are promising future directions that could further revolutionize these fields.

1.7.1 Technical Challenges

As the integration of AI, optical, and wireless communication technologies continues to evolve, several technical challenges must be addressed to ensure the successful deployment and operation of these systems.

1.7.1.1 Data Availability and Quality

One of the significant challenges in applying AI to optical and wireless communication is the difficulty in obtaining and ensuring high-quality data necessary for training ML models. Robust datasets are essential for building AI models that can make accurate predictions and informed decisions. However, acquiring large, diverse, and well-labeled datasets, particularly for infrequent events like network failures or security breaches, can be quite challenging.

To overcome this obstacle, researchers are exploring methods such as data augmentation, transfer learning, and synthetic data generation. These strategies can help create more comprehensive datasets, which are crucial for developing more effective and reliable AI models.

1.7.1.2 Computational Complexity and Power Efficiency

AI algorithms, particularly deep learning models, are often computationally intensive, requiring significant processing power and energy. In resource-constrained environments, such as mobile devices or edge computing nodes, this computational complexity can be a limiting factor.

To address this issue, there is ongoing research into developing more efficient AI models and algorithms that can operate within these constraints. Techniques such as model compression, quantization, and edge AI (where computation is distributed closer to the data source) are being explored to reduce the computational burden and improve energy efficiency.

1.7.1.3 Interoperability and Standardization

As optical and wireless networks become more integrated, ensuring interoperability between different technologies and standards is a significant challenge. The lack of standardized protocols and interfaces can lead to compatibility issues, hindering the seamless operation of integrated networks.

Standardization efforts are critical to overcoming this challenge. Industry bodies and research organizations are working on developing unified frameworks and standards that can facilitate the integration of optical and wireless technologies. Additionally, AI-driven solutions can help manage and optimize interoperability, dynamically adapting to different network conditions and configurations.

1.7.1.4 Security and Privacy

The integration of AI into communication networks introduces new security and privacy concerns. AI systems, particularly those based on ML, can be vulnerable to attacks such as adversarial ML, where malicious actors manipulate input data to deceive the model.

Ensuring the security and privacy of AI-driven communication networks requires robust defense mechanisms. This includes developing secure AI models, implementing encryption and authentication protocols, and continuously monitoring for potential threats. Moreover, AI itself can be used to enhance network security by detecting and mitigating attacks in real-time.

1.7.2 Ethical and Privacy Considerations

As AI becomes more deeply integrated into communication networks, ethical and privacy considerations are becoming increasingly important. The use of AI to manage and optimize communication networks raises questions about data privacy, transparency, and accountability.

1.7.2.1 Data Privacy

People use AI systems in various applications where big data is common; this has raised some concerns on how the data is obtained and used. Security is a main concern when it comes to data privacy, especially information exchanged over different communication channels.

In order to mitigate these issues, it is necessary to adopt severe data governance measures and procedures. This includes elimination of personal identification data, proper storage and protection of data, and user control of the data. To prevent the misuse of user data, AI privacy-preserving methods, including differential privacy and federated learning, among others are being considered as desirable.

1.7.2.2 Transparency and Explainability

One of the known facts about AI models, especially deep learning models, is that they are opaque, or in other words – are "black boxes"; their internal decisions are not easily explainable. In the areas which require decision-making, it is crucial to make the decision of AI systems explainable in the terms of the human language.

Building XAI or "transparent and explainable AI" is an area of ongoing research and aimed at designing models capable of justifying their actions or choices. Such practices are important for establishing trust in the AI systems and making sure that such systems are applied appropriately.

1.7.2.3 Ethical AI Use

The ethical use of AI in communication networks involves ensuring that AI systems are designed and deployed in a way that is fair, unbiased, and does not harm users. This includes avoiding algorithmic biases that could lead to discriminatory outcomes, ensuring that AI systems are used to enhance rather than replace human decision-making and considering the broader societal impacts of AI deployment.

Developing ethical guidelines and frameworks for AI use in communication networks is essential for addressing these concerns. Collaboration between industry, academia, and policymakers is crucial to ensure that AI is used responsibly and for the benefit of all users.

1.7.3 Future Research Directions

Looking forward, there are several promising research directions in the fields of optical and wireless communication, particularly concerning the integration of AI.

1.7.3.1 AI-Driven Network Automation

As networks become more complex, there is a growing need for automation in network management and operation. AI-driven automation can help manage the complexity of modern networks, enabling more efficient resource allocation, fault detection, and performance optimization. Future research will likely focus on developing more sophisticated AI models that can autonomously manage networks, reducing the need for human intervention and improving overall network performance.

1.7.3.2 6G and Beyond

While 5G networks are still being deployed, research into 6G networks is already underway. 6G is expected to bring even higher data rates, lower latency, and more reliable connectivity, supporting applications such as holographic communication, immersive augmented reality, and advanced IoT.

AI will play a central role in 6G, enabling new communication paradigms and optimizing network performance. Research in this area will likely focus on integrating AI more deeply into the core of communication networks, exploring new spectrum bands (such as terahertz frequencies) and developing novel communication techniques that leverage AI.

1.7.3.3 Quantum Communication

Quantum communication is an emerging field that has the potential to revolutionize secure communication networks. By leveraging the principles of quantum mechanics, quantum communication can provide theoretically unbreakable security, making it an attractive option for applications requiring high levels of data protection.

The integration of AI with quantum communication is an exciting area of research. AI can help optimize quantum communication protocols, manage quantum networks, and develop new applications that leverage the unique properties of quantum communication.

1.7.3.4 AI for Environmental Sustainability

Since issues of the environmental cost of communication networks are increasingly being addressed, the emphasis is placed on creating environmentally friendly technologies. Energy efficiency also poses the opportunity in future to assist this process through focusing on power utilization, its decrease, and more rational consumption of networks' capacities.

Future work will probably investigate AI-assisted green communication networks, known to cover topics such as utilizing energy efficient data centers, low power wireless networks, and green communications technology.

1.8 CONCLUSION

The rapid advancements in optical and wireless communication, fueled by the integration of ML and AI, have ushered in a new era of connectivity and technological innovation. Throughout this chapter, we have explored the fundamental aspects of these communication technologies, the transformative role of AI, the state of the art, and the challenges and future directions that lie ahead.

1.8.1 Summary of Key Points

The convergence of optical and wireless communication technologies is reshaping the landscape of global networks. Optical communication, with its unparalleled capacity and efficiency, forms the backbone of modern telecommunication systems, while wireless communication provides the flexibility and accessibility required for last-mile connectivity. The integration of these two domains, particularly through the application of AI, has enabled the creation of more dynamic, efficient, and intelligent networks.

We began by exploring the fundamentals of optical and wireless communication, highlighting the key technologies, principles, and challenges in each domain. The discussion then shifted to the role of AI in enhancing these communication systems, from optimizing signal processing and network management to improving security and fault detection. AI's ability to learn from data, adapt to changing conditions, and make predictive decisions has opened new possibilities for both optical and wireless networks, making them more robust, efficient, and capable of meeting the demands of modern applications.

The chapter also delved into the state of the art in optical and wireless communication, showcasing the latest advancements in areas such as coherent optical communication, space-division multiplexing, 5G technologies, and AI-enhanced wireless networks. These innovations are driving the evolution of communication networks, enabling faster data transmission, lower latency, and more reliable connectivity.

However, with these advancements come significant challenges. The technical complexities of integrating AI with communication networks, the need for large and high-quality datasets, computational efficiency, interoperability, and security are all critical issues that must be addressed. Moreover, the ethical and privacy implications of AI in communication systems require careful consideration, ensuring that these technologies are developed and deployed responsibly.

Finally, we explored the future directions in this rapidly evolving field. From the development of 6G networks and quantum communication to AI-driven network automation and environmental sustainability, the future of optical and wireless communication holds immense promise. AI will continue to play a central role in this evolution, driving new innovations and transforming how we connect and communicate.

1.8.2 The Road Ahead

As we look ahead, the fusion of AI with optical and wireless communication technologies is expected to be a significant catalyst for the evolution of next-generation communication networks. The emergence of 6G networks, which will prioritize ultra-reliable, low-latency communication, massive connectivity, and enhanced mobile broadband, is set to push the limits of what is possible in wireless communications. AI will play a central role in this transformation, enabling innovations such as intelligent resource management, dynamic spectrum allocation, and real-time network optimization.

These technological advancements are likely to have far-reaching societal implications, fundamentally changing the way we live, work, and interact. For instance, AI-driven communication networks could lead to more efficient and adaptive urban infrastructures, expand global connectivity in underserved areas, and support new applications in fields like healthcare, education, and disaster management.

To fully harness the potential of these innovations, collaboration between industry, academia, and policymakers will be crucial. Developing standardized frameworks, ethical guidelines, and robust security measures will be essential to ensuring that AI-driven communication networks are safe, reliable, and beneficial for everyone.

In summary, the integration of AI with optical and wireless communication technologies is set to reshape the future of global connectivity. While the challenges are substantial, the opportunities are even greater. As we continue to innovate and explore new horizons, the vision of an intelligent and interconnected world is becoming increasingly attainable, powered by the ongoing advancements in AI, optical, and wireless communication technologies.

REFERENCES

[1] K. Zeb, "Quantum Dash Multi-Wavelength Lasers for Next Generation High Capacity Multi-Gb/s Millimeter-Wave Radio-over-Fiber Wireless Communication Networks," PhD Thesis, Concordia University, 2023.

[2] A. Sharma and V. Mishra, "Enhancing high-speed networks using RGB-based WLAN through Ro-FSO integration in the 5 GHz band," *Journal of Optical Communications,* 2023.

[3] A. Sharma, V. Mishra, K. Singh, and J. Malhotra, "Hybrid RoF-RoFSO system for broadband services by incorporating polarization division multiplexing scheme," *Journal of Optical Communications,* 2023.

[4] A. Sharma, K. Singh, and J. Malhotra, "High speed 60 Gbps RGB laser based-FSOC link by incorporating hybrid PDM-MIMO scheme for indoor applications," *Journal of Optical Communications,* 2023.

[5] N. J. Gomes, P. P. Monteiro, and A. Gameiro, *Next Generation Wireless Communications Using Radio over Fiber*: Hoboken: John Wiley & Sons, 2012.

[6] W. Aman, S. Al-Kuwari, M. Muzzammil, M. M. U. Rahman, and A. Kumar, "Security of underwater and air–water wireless communication: State-of-the-art, challenges and outlook," *Ad Hoc Networks,* vol. 142, p. 103114, 2023.

[7] Z. Qadir, K. N. Le, N. Saeed, and H. S. Munawar, "Towards 6G Internet of Things: Recent advances, use cases, and open challenges," *ICT Express,* vol. 9, pp. 296–312, 2023.

[8] W. Xiong, P. Ambichl, Y. Bromberg, B. Redding, S. Rotter, and H. Cao, "Spatiotemporal control of light transmission through a multimode fiber with strong mode coupling," *Physical Review Letters,* vol. 117, p. 053901, 2016.

[9] C. Ioannou and V. Vassiliou, "Network attack classification in IoT using support vector machines," *Journal of Sensor and Actuator Networks,* vol. 10, p. 58, 2021.

[10] S. Yang, P. Liu, and T. H. Lim, "IoT-based underwater robotics for water quality monitoring in aquaculture: A survey," in *The 11th International Conference on Robot Intelligence Technology and Applications (RiTA 2023)*, Taicang, China, 2023.

[11] M. Maurya, I. Panigrahi, D. Dash, and C. Malla, "Intelligent fault diagnostic system for rotating machinery based on IoT with cloud

computing and artificial intelligence techniques: A review," *Soft Computing*, vol. 28, pp. 477–494, 2024.

[12] C. Zhang, P. Liang, J. Nebhen, S. Chaudhary, A. Sharma, J. Malhotra, *et al.*, "Performance analysis of mode division multiplexing-based free space optical systems for healthcare infrastructure's," *Optical and Quantum Electronics*, vol. 53, pp. 1–14, 2021.

[13] A. Sharma and R. Kapoor, "Study of various challenges in is OWC: A review," *International Journal for Research in Applied Science and Engineering Technology*, vol. 5, pp. 802–807, 2017.

[14] A. Sharma and D. Thakur, "A review on WLANs with radio-over-fiber technology," *The International Journal of Electronics, Communications, and Engineering (IJECE)*, vol. 6, pp. 1–6, 2017.

[15] Z. Zhou, H. Zhang, C. Lin, and A. Sharma, "Performance analysis of duobinary and CSRZ modulation based polarization interleaving for high-speed WDM-FSO transmission system," *Journal of Optical Communications*, vol. 43, pp. 147–152, 2022.

[16] S. Chaudhary, A. Sharma, and N. Chaudhary, "6 × 20 Gbps hybrid WDM–PI inter-satellite system under the influence of transmitting pointing errors," *Journal of Optical Communications*, vol. 37, pp. 375–379, 2016.

[17] S. Chaudhary and A. Sharma, "6 x 20Gbps long reach WDM-PI based high altitude platform inter-satellite communication system," *International Journal of Computer Applications*, vol. 122, pp. 41–45, 2015.

[18] S. Chaudhary, P. Chauhan, and A. Sharma, "High speed 4 × 2.5 Gbps-5 GHz AMI-WDM-RoF transmission system for WLANs," *Journal of Optical Communications*, vol. 40, pp. 285–288, 2019.

[19] S. Chaudhary, D. Thakur, and A. Sharma, "10 Gbps-60 GHz RoF transmission system for 5 G applications," *Journal of Optical Communications*, vol. 40, pp. 281–284, 2019.

[20] A. Sharma, S. Chaudhary, D. Thakur, and V. Dhasratan, "A cost-effective high-speed radio over fibre system for millimeter wave applications," *Journal of Optical Communications*, vol. 41, pp. 177–180, 2020.

[21] N. Vishwakarma, R. Swaminathan, P. D. Diamantoulakis, and G. K. Karagiannidis, "Performance analysis of optical reflecting surface-assisted optical space shift keying-based MIMO-FSO system," *IEEE Transactions on Communications*, vol. 71, pp. 4751–4763, 2023.

[22] A. Sood and R. Kaushik, "4× 20 Gbps OFDM based RoFSO transmission using hybrid SCM-MDM under different turbulence conditions," *Advances in Electrical and Electronic Engineering,* vol. 21, pp. 48–58, 2023.

[23] M. Singh, M. H. Aly, and S. A. Abd El-Mottaleb, "6G enabling FSO communication system employing integrated PDM-OAM-OCDMA transmission: Impact of weather conditions in India," *Applied Optics,* vol. 62, pp. 142–152, 2023.

[24] S. Khichar, P. Sasithong, H. L. Aung, A. Sharma, S. Chaudhary, W. Santipach, *et al.*, "MIMO-NOMA With mmWave transmission," in *Applications of 5G and Beyond in Smart Cities,* ed: A. Bajpai, A. Balodi, Boca Raton: CRC Press, 2023, pp. 153–168.

[25] J. B. Horst, *MIMO Signals and Systems*: New York: Springer Science-HBusiness Media, Inc, 2005.

[26] N. B. Sinha, M. C. Snai, and M. Mitra, "Performance enhancement of MIMO-OFDM technology for high data rate wireless networks," *International Journal of Computer Science and Application,* vol. 2010, pp. 122–128, 2010.

[27] H. V. Balan, R. Rogalin, A. Michaloliakos, K. Psounis, and G. Caire, "Achieving high data rates in a distributed MIMO system," in *Proceedings of the 18th Annual International Conference on Mobile Computing and Networking,* Istanbul, Turkey, 2012, pp. 41–52.

[28] M. Singh, M. Lal Singh, R. Singh, H. Kaur, and P. Chohan, "Mitigation of turbulences induced by air bubbles and performance enhancement in 112 Gbps UWOC link with coherent detection MIMO PDM-QPSK and advanced digital signal processing," *Optical and Quantum Electronics,* vol. 54, p. 472, 2022.

[29] M. A. S. Sejan, M. H. Rahman, M. A. Aziz, D.-S. Kim, Y.-H. You, and H.-K. Song, "A comprehensive survey on MIMO visible light communication: Current research, machine learning and future trends," *Sensors,* vol. 23, p. 739, 2023.

[30] S. Sinha, C. Kumar, A. Armghan, M. Singh, M. Alsharari, and K. Aliqab, "Simulative investigation of the hybrid, spatially multiplexed MIMO-FSO transmission system under atmospheric turbulence," *IEEE Access,* vol. 11, pp. 55071–55080, 2023.

[31] Z. Yang, Y. Fang, L. Lv, X. Yu, and Y. L. Guan, "A low-latency iterative decoding architecture for MIMO-VLC systems with GSM," in

2023 IEEE/CIC International Conference on Communications in China (ICCC Workshops), Dalian, China, 2023, pp. 1–6.

[32] L. Li and A. Sharma, "High speed RGB-based duobinary-encoded visible light communication system under the impact of turbulences," *Frontiers in Physics*, vol. 10, p. 944623, 2022.

[33] A. Sharma, K. Singh, and J. Malhotra, "High speed 60 Gbps RGB laser based-FSOC link by incorporating hybrid PDM-MIMO scheme for indoor applications," *Journal of Optical Communications*, 2023.

[34] A. Tripathi, S. Gupta, and A. Mandloi, "Orthogonally polarized and 60 GHz dual-channel based 18 × 2.5 Gb/s DWDM-interleaved hybrid FSO system under atmospheric turbulence," *Optical and Quantum Electronics*, vol. 52, p. 220, 2020.

[35] K. H. Shakthi Murugan, A. Sharma, and J. Malhotra, "Performance analysis of 80 Gbps Ro-FSO system by incorporating hybrid WDM-MDM scheme," *Optical and Quantum Electronics*, vol. 52, p. 505, 2020.

[36] I. B. Kovacs, M. S. Faruk, and S. J. Savory, "200 Gb/s/λ upstream PON using polarization multiplexed PAM4 with coherent detection," *IEEE Photonics Technology Letters*, vol. 35, pp. 1014–1017, 2023.

[37] P. Mandal, N. Sarkar, S. Santra, B. Dutta, B. Kuiri, K. Mallick, *et al.*, "Hybrid WDM-FSO-PON with integrated SMF/FSO link for transportation of Rayleigh backscattering noise mitigated wired/wireless information in long-reach," *Optics Communications*, vol. 507, p. 127594, 2022.

[38] E. E. Elsayed, D. Kakati, M. Singh, A. Grover, and G. Anand, "Design and analysis of a dense wavelength-division multiplexed integrated PON-FSO system using modified OOK/DPPM modulation schemes over atmospheric turbulences," *Optical and Quantum Electronics*, vol. 54, p. 768, 2022.

[39] N. Chi, M. Zhang, Y. Zhou, and J. Zhao, "3.375-Gb/s RGB-LED based WDM visible light communication system employing PAM-8 modulation with phase shifted Manchester coding," *Optics Express*, vol. 24, pp. 21663–21673, 2016.

[40] S. Chaudhary, A. Sharma, Q. Li, Y. Meng, and J. Malhotra, "Enhancing sustainable transportation with advancements in photonic radar technology with MIMO and IIR filtering for adverse weather conditions," *Sustainability*, vol. 16, p. 5426, 2024.

[41] S. Taba, "Contribution to the Development of Advanced Techniques for Massive MIMO Communication Systems," Doctoral Thesis, Guelma University, 2023.

[42] S. Sun, F. Yang, J. Song, and R. Zhang, "Intelligent reflecting surface for MIMO VLC: Joint design of surface configuration and transceiver signal processing," *IEEE Transactions on Wireless Communications,* vol. 22, pp. 5785–5799, 2023.

[43] S. Hamid, S. R. Chopra, A. Gupta, S. Tanwar, B. C. Florea, D. D. Taralunga, *et al.*, "Hybrid beamforming in massive MIMO for next-generation communication technology," *Sensors,* vol. 23, p. 7294, 2023.

[44] K. S. Mohamed, M. Y. Alias, M. Roslee, and Y. M. Raji, "Towards green communication in 5G systems: Survey on beamforming concept," *IET Communications,* vol. 15, pp. 142–154, 2021.

[45] J. Zhang, X. Yu, and K. B. Letaief, "Hybrid beamforming for 5G and beyond millimeter-wave systems: A holistic view," *IEEE Open Journal of the Communications Society,* vol. 1, pp. 77–91, 2020.

[46] M. R. Mahmood, M. A. Matin, P. Sarigiannidis, and S. K. Goudos, "A comprehensive review on artificial intelligence/machine learning algorithms for empowering the future IoT toward 6G era," *IEEE Access,* vol. 10, pp. 87535–87562, 2022.

[47] S. S. Gill, M. Xu, C. Ottaviani, P. Patros, R. Bahsoon, A. Shaghaghi, *et al.*, "AI for next generation computing: Emerging trends and future directions," *Internet of Things,* vol. 19, p. 100514, 2022.

[48] S. Chaudhary, A. Sharma, K. Singh, S. Khichar, and J. Malhotra, "Highly efficient photonic radar by incorporating MDM-WDM and machine learning classifiers under adverse weather conditions," *Plos One,* vol. 19, p. e0300653, 2024.

[49] A. Sharma, J. Malhotra, S. Khichar, A. Parnianifard, and S. Chaudhary, "Highly efficient frequency modulated continuous wave based photonic radar by incorporating electronic equalization scheme," *Optical and Quantum Electronics,* vol. 55, p. 797, 2023.

[50] E. Reyes-Vera, A. Valencia-Arias, V. García-Pineda, E. F. Aurora-Vigo, H. Alvarez Vásquez, and G. Sánchez, "Machine learning applications in optical fiber sensing: A research agenda," *Sensors,* vol. 24, p. 2200, 2024.

[51] Y. Zhou, Y.-N. Zhang, Q. Yu, L. Ren, Q. Liu, and Y. Zhao, "Application of machine learning in optical fiber sensors," *Measurement*, vol. 228, p. 114391, 2024.

[52] S. Chaudhary, A. Sharma, S. Khichar, Y. Meng, and J. Malhotra, "Enhancing autonomous vehicle navigation using SVM-based multi-target detection with photonic radar in complex traffic scenarios," *Scientific Reports*, vol. 14, p. 17339, 2024.

[53] S. A. Saad, I. Shayea, and N. M. O. Sid Ahmed, "Artificial intelligence linear regression model for mobility robustness optimization algorithm in 5G cellular networks," *Alexandria Engineering Journal*, vol. 89, pp. 125–148, 2024.

[54] R. Chataut, M. Nankya, and R. Akl, "6G networks and the AI revolution—Exploring technologies, applications, and emerging challenges," *Sensors*, vol. 24, p. 1888, 2024.

[55] A. Karapantelakis, P. Alizadeh, A. Alabassi, K. Dey, and A. Nikou, "Generative AI in mobile networks: a survey," *Annals of Telecommunications*, vol. 79, pp. 15–33, 2024.

[56] S. Mahboob and L. Liu, "Revolutionizing future connectivity: A contemporary survey on AI-empowered satellite-based non-terrestrial networks in 6G," *IEEE Communications Surveys & Tutorials*, vol. 26, pp. 1279–1321, 2024.

[57] A. R. Sahu, S. K. Palei, and A. Mishra, "Data-driven fault diagnosis approaches for industrial equipment: A review," *Expert Systems*, vol. 41, p. e13360, 2024.

[58] A. Abid, M. T. Khan, and J. Iqbal, "A review on fault detection and diagnosis techniques: Basics and beyond," *Artificial Intelligence Review*, vol. 54, pp. 3639–3664, 2021.

[59] M. C. Hlophe and B. T. Maharaj, "AI meets CRNs: A prospective review on the application of deep architectures in spectrum management," *IEEE Access*, vol. 9, pp. 113954–113996, 2021.

[60] R. D. Apaza, E. J. Knoblock, and H. Li, "A new spectrum management concept for future NAS communications," in *2020 AIAA/IEEE 39th Digital Avionics Systems Conference (DASC)*, San Antonio, TX, 2020, pp. 1–7.

[61] Y. C. Liang, J. Tan, H. Jia, J. Zhang, and L. Zhao, "Realizing intelligent spectrum management for integrated satellite and terrestrial networks," *Journal of Communications and Information Networks*, vol. 6, pp. 32–43, 2021.

[62] K. Thakkar, A. Goyal, and B. Bhattacharyya, "Deep learning and channel estimation," in *2020 6th International Conference on Advanced Computing and Communication Systems (ICACCS)*, Coimbatore, India, 2020, pp. 745–751.

[63] E. C. Vilas Boas, J. D. S. de Silva, F. A. P. de Figueiredo, L. L. Mendes, and R. A. A. de Souza, "Artificial intelligence for channel estimation in multicarrier systems for B5G/6G communications: A survey," *EURASIP Journal on Wireless Communications and Networking*, vol. 2022, p. 116, 2022.

[64] A. K. Gizzini, Y. Medjahdi, A. J. Ghandour, and L. Clavier, "Towards explainable AI for channel estimation in wireless communications," *IEEE Transactions on Vehicular Technology*, vol. 73, pp. 7389–7394, 2024.

[65] Y. Jin, J. Zhang, S. Jin, and B. Ai, "Channel estimation for cell-free mmWave massive MIMO through deep learning," *IEEE Transactions on Vehicular Technology*, vol. 68, pp. 10325–10329, 2019.

[66] A. S. Jacobs, R. Beltiukov, W. Willinger, R. A. Ferreira, A. Gupta, and L. Z. Granville, "AI/ML for network security: The emperor has no clothes," Presented at the *Proceedings of the 2022 ACM SIGSAC Conference on Computer and Communications Security*, Los Angeles, CA, 2022.

[67] Y. Siriwardhana, P. Porambage, M. Liyanage, and M. Ylianttila, "AI and 6G security: Opportunities and challenges," in *2021 Joint European Conference on Networks and Communications & 6G Summit (EuCNC/6G Summit)*, Porto, Portugal, 2021, pp. 616–621.

[68] E. Bertino, M. Kantarcioglu, C. G. Akcora, S. Samtani, S. Mittal, and M. Gupta, "AI for security and security for AI," Presented at the *Proceedings of the Eleventh ACM Conference on Data and Application Security and Privacy*, Virtual Event, USA, 2021.

[69] M. Waqas, S. Tu, Z. Halim, S. U. Rehman, G. Abbas, and Z. H. Abbas, "The role of artificial intelligence and machine learning in wireless networks security: principle, practice and challenges," *Artificial Intelligence Review*, vol. 55, pp. 5215–5261, 2022.

[70] L. Xiao, X. Wan, X. Lu, Y. Zhang, and D. Wu, "IoT security techniques based on machine learning: How do IoT devices use AI to enhance security?" *IEEE Signal Processing Magazine*, vol. 35, pp. 41–49, 2018.

[71] T. Maksymyuk, J. Gazda, O. Yaremko, and D. Nevinskiy, "Deep learning based massive MIMO beamforming for 5G mobile network," in *2018 IEEE 4th International Symposium on Wireless Systems within the International Conferences on Intelligent Data Acquisition and Advanced Computing Systems (IDAACS-SWS)*, Lviv, Ukraine, 2018, pp. 241–244.

[72] E. Ali, M. Ismail, R. Nordin, and N. F. Abdulah, "Beamforming techniques for massive MIMO systems in 5G: Overview, classification, and trends for future research," *Frontiers of Information Technology & Electronic Engineering*, vol. 18, pp. 753–772, 2017.

[73] M. Kanaka Chary, C. H. Vamshi Krishna, and D. Rama Krishna, "Accurate channel estimation and hybrid beamforming using Artificial Intelligence for massive MIMO 5G systems," *AEU – International Journal of Electronics and Communications*, vol. 173, p. 154971, 2024.

[74] S. Chaudhary and A. Amphawan, "Optimization of AMI-MDM-RoFSO under atmospheric turbulence," *EPJ Web Conferences*, vol. 162, p. 01020, 2017.

[75] A. Amphawan, S. Chaudhary, R. Din, and M. N. Omar, "5Gbps HG 0, 1 and HG 0, 3 optical mode division multiplexing for RoFSO," in *2015 IEEE 11th International Colloquium on Signal Processing & Its Applications (CSPA)*, Kuala Lumpur, Malaysia, 2015, pp. 145–149.

[76] S. Chaudhary, A. Sharma, M. A. Naeem, and Y. Meng, "Target detection in challenging environments: Photonic radar with a hybrid multiplexing scheme for 5G autonomous vehicles," *Sustainability*, vol. 16, p. 991, 2024.

[77] R.-J. Essiambre and R. W. Tkach, "Capacity trends and limits of optical communication networks," *Proceedings of the IEEE*, vol. 100, pp. 1035–1055, 2012.

[78] H. Kaushal and G. Kaddoum, "Optical communication in space: Challenges and mitigation techniques," *IEEE Communications Surveys & Tutorials*, vol. 19, pp. 57–96, 2016.

[79] L. Song, X. Hu, G. Zhang, P. Spachos, K. N. Plataniotis, and H. Wu, "Networking systems of AI: On the convergence of computing and communications," *IEEE Internet of Things Journal*, vol. 9, pp. 20352–20381, 2022.

[80] J. Mata, I. de Miguel, R. J. Durán, N. Merayo, S. K. Singh, A. Jukan, *et al.*, "Artificial intelligence (AI) methods in optical networks: A

comprehensive survey," *Optical Switching and Networking*, vol. 28, pp. 43–57, 2018.
[81] A. Luusua, J. Ylipulli, M. Foth, and A. Aurigi, "Urban AI: Understanding the emerging role of artificial intelligence in smart cities," *AI & Society*, vol. 38, pp. 1039–1044, 2023.
[82] M. E. Zamponi and E. Barbierato, "The dual role of artificial intelligence in developing smart cities," *Smart Cities*, vol. 5, pp. 728–755, 2022.
[83] I. P. Chochliouros, N. Yi, A. S. Spiliopoulou, A. Kostopoulos, N. Gomes, U. Herzog, *et al.*, "Enhanced mobile broadband as enabler for 5G: Actions from the framework of the 5G-DRIVE project," in *Artificial Intelligence Applications and Innovations, AIAI 2019: MHDW and 5G-PINE 2019, Proceedings*, ed: I. Maglogiannis, L. Iliadis, J. MacIntyre, and E. Pimenidis, Cham: Springer, 2019, pp. 31–45.
[84] L. Zhang, Y. C. Liang, and D. Niyato, "6G visions: Mobile ultra-broadband, super internet-of-things, and artificial intelligence," *China Communications*, vol. 16, pp. 1–14, 2019.
[85] A. Salh, L. Audah, N. S. M. Shah, A. Alhammadi, Q. Abdullah, Y. H. Kim, *et al.*, "A survey on deep learning for ultra-reliable and low-latency communications challenges on 6G wireless systems," *IEEE Access*, vol. 9, pp. 55098–55131, 2021.
[86] J. Park, S. Samarakoon, H. Shiri, M. K. Abdel-Aziz, T. Nishio, A. Elgabli, *et al.*, "Extreme ultra-reliable and low-latency communication," *Nature Electronics*, vol. 5, pp. 133–141, 2022.
[87] A. Sharma and V. Kumar, "High speed and long reach DPSK-OFDM-Is-OWC under impact of space turbulences," *International Journal of Computer Applications*, vol. 975, p. 8887, 2018.
[88] A. Sharma and R. Kapoor, "Study of various challenges in is OWC: A review," *Journal for Research in Applied Science and Engineering Technology (IJRASET)*, vol. 5, pp. 802–807, 2017.
[89] S. Chaudhary, L. Wuttisittikulkij, J. Nebhen, A. Sharma, D. Z. Rodriguez, and S. Kumar, "Terabyte capacity-enabled (10 × 400 Gbps) Is-OWC system for long-haul communication by incorporating dual polarization quadrature phase shift key and mode division multiplexing scheme," *PLoS One*, vol. 17, p. e0265044, 2022.
[90] X. Zhou and C. Xie, *Enabling Technologies for High Spectral-Efficiency Coherent Optical Communication Networks*: Hoboken: John Wiley & Sons, 2016.

[91] J. Zhang and Z. Jia, "Coherent passive optical networks for 100G/λ-and-beyond fiber access: Recent progress and outlook," *IEEE Network*, vol. 36, pp. 116–123, 2022.
[92] X. Tang, Z. Ghassemlooy, S. Rajbhandari, W. O. Popoola, and C. G. Lee, "Coherent heterodyne multilevel polarization shift keying with spatial diversity in a free-space optical turbulence channel," *Journal of Lightwave Technology*, vol. 30, pp. 2689–2695, 2012.
[93] V. Sharma, M. Lumba, and G. Kaur, "Severe climate sway in coherent CDMA-OSSB-FSO transmission system," *Optik*, vol. 125, pp. 5705–5707, 2014.
[94] R. Ryf, S. Randel, A. H. Gnauck, C. Bolle, R.-J. Essiambre, P. Winzer, *et al.*, "Space-division multiplexing over 10 km of three-mode fiber using coherent 6 × 6 MIMO processing," in *National Fiber Optic Engineers Conference*, Los Angeles, CA, 2011, p. PDPB10.
[95] V. Samuthira Pandi, M. Singh, A. Grover, J. Malhotra, and S. Singh, "Performance analysis of 400 Gbit/s hybrid space division multiplexing-polarization division multiplexing-coherent detection-orthogonal frequency division multiplexing-based free-space optics transmission system," *International Journal of Communication Systems*, vol. 35, p. e5310, 2022.
[96] T. M. Olaleye, P. A. Ribeiro, and M. Raposo, "Generation of photon orbital angular momentum and its application in space division multiplexing," *Photonics*, vol. 10, p. 664, 2023.
[97] A. Amphawan, S. Chaudhary, T.-K. Neo, M. Kakavand, and M. Dabbagh, "Radio-over-free space optical space division multiplexing system using 3-core photonic crystal fiber mode group multiplexers," *Wireless Networks*, vol. 27, pp. 211–225, 2021.
[98] W. A. Imtiaz, H. Y. Ahmed, M. Zeghid, and Y. Sharief, "An optimized architecture to reduce the impact of fiber strands in spectral/spatial optical code division multiple access passive optical networks (OCDMA-PON)," *Optical Fiber Technology*, vol. 54, p. 102072, 2020.
[99] W. A. Imtiaz, H. Y. Ahmed, M. Zeghid, and Y. Sharief, "Two dimensional optimized enhanced multi diagonal code for OCDMA passive optical networks," *Optical and Quantum Electronics*, vol. 52, p. 33, 2019/12/13 2019.

[100] M. Singh, J. Kříž, M. Kamruzzaman, V. Dhasarathan, A. Sharma, and S. A. Abd El-Mottaleb, "Design of a high-speed OFDM-SAC-OCDMA-based FSO system using EDW codes for supporting 5G data services and smart city applications," *Frontiers in Physics,* vol. 10, p. 934848, 2022.

[101] A. Amphawan, S. Chaudhary, and V. Chan, "2 × 20 Gbps-40 GHz OFDM Ro-FSO transmission with mode division multiplexing," *Journal of the European Optical Society-Rapid publications,* vol. 9, 2014.

[102] H. Sarangal, A. Singh, J. Malhotra, and S. Chaudhary, "A cost effective 100 Gbps hybrid MDM–OCDMA–FSO transmission system under atmospheric turbulences," *Optical and Quantum Electronics,* vol. 49, p. 184, 2017.

[103] A. Sharma and J. Malhotra, "Evaluating the effects of material reflectivity and atmospheric attenuation on photonic radar performance in free space optical channels," *Journal of Optical Communications,* 2023.

[104] S. Chaudhary, P. Chauhan, and A. Sharma, "High speed 4 × 2.5 Gbps-5 GHz AMI-WDM-RoF transmission system for WLANs," *Journal of Optical Communications.,* vol. 40, pp. 285–288, 2019.

[105] Z. Zhou, H. Zhang, C. Lin, and A. Sharma, "Performance analysis of duobinary and CSRZ modulation based polarization interleaving for high-speed WDM-FSO transmission system," *Journal of Optical Communications,* vol. 43, pp. 147–152, 2018.

[106] P. Sood and A. Sharma, "Chandni. Analysis of FSO system and its challenges–A review," *International Journal of Computer Applications,* vol. 179, pp. 42–45, 2018.

[107] A. Sharma and R. Kapoor, "Study of various challenges in is OWC: A review," *International Journal for Research in Applied Science and Engineering Technology (IJRASET),* vol. 5, pp. 802–807, 2017.

[108] A. Sharma and S. Rana, "Implementation of radio over fiber technology with different filtration techniques." *International Journal for Research in Applied Science & Engineering Technology (IJRASET),* vol. 5, pp. 783–789, 2017.

[109] A. Sharma, V. Kumar, and V. Gupta, "A review on inter-satellite optical wireless communication," *International Journal of Computer Applications,* vol. 180, pp. 13–17, 2018.

[110] A. Sharma and P. Chauhan, “High speed radio over fiber system for wireless local area networks by incorporating alternate mark inversion scheme.” *Journal of Optical Communications*, vol. 42, p. 273, 2021.

[111] S. Chaudhary, D. Thakur, and A. Sharma, “10 Gbps-60 GHz RoF transmission system for 5 G applications.” *Journal of Optical Communications*, vol. 40, p. 281, 2019.

[112] A. Sharma and J. Malhota, “Performance investigation of photonic radar for autonomous vehicles’ application under various degrading conditions,” in *Signals, Machines and Automation*, ed: A. Rani, B. Kumar, V. Shrivastava, and R. C. Bansal, Singapore: Springer, 2023, pp. 505–512.

[113] S. Chaudhary, A. Sharma, S. Khichar, S. Shah, R. Ullah, A. Parnianifard, *et al.*, “A salinity-impact analysis of polarization division multiplexing-based underwater optical wireless communication system with high-speed data transmission,” *Journal of Sensor and Actuator Networks*, vol. 12, p. 72, 2023.

[114] A. Sharma and J. Malhotra, “Performance enhancement of photonic radar sensor for detecting multiple targets by incorporating mode division multiplexing,” *Optical and Quantum Electronics*, vol. 54, p. 410, 2022.

[115] A. Sharma and J. Malhotra, “Simulative investigation of FMCW based optical photonic radar and its different configurations,” *Optical and Quantum Electronics*, vol. 54, p. 233, 2022.

[116] A. Sharma, S. Chaudhary, J. Malhotra, M. Saadi, S. Al Otaibi, and L. Wuttisittikulkij, “Speed-direction sensing under multiple vehicles scenario using photonic radars,” *CMC-Computers Materials & Continua*, vol. 73, pp. 5399–5410, 2022.

[117] A. Sharma, S. Chaudhary, J. Malhotra, A. Parnianifard, and L. Wuttisittikulkij, “Measurement of target range and doppler shift by incorporating PDM-enabled FMCW-based photonic radar,” *Optik*, vol. 262, p. 169191, 2022.

[118] A. Sharma, S. Chaudhary, J. Malhotra, S. Khichar, and L. Wuttisittikulkij, “Photonic sensor for multiple targets detection under adverse weather conditions in autonomous vehicles,” *Journal of Sensor and Actuator Networks*, vol. 11, p. 60, 2022.

[119] S. Chaudhary, A. Sharma, S. Khichar, X. Tang, X. Wei, and L. Wuttisittikulkij, "High resolution-based coherent photonic radar sensor for multiple target detections," *Journal of Sensor and Actuator Networks,* vol. 11, p. 49, 2022.

[120] A. Sharma, J. Malhotra, S. Chaudhary, and V. Thappa, "Analysis of 2 × 10 Gbps MDM enabled inter satellite optical wireless communication under the impact of pointing errors," *Optik,* vol. 227, p. 165250, 2021.

[121] A. Sharma, S. Chaudhary, J. Malhotra, M. Saadi, S. Al Otaibi, J. Nebhen, *et al.*, "A cost-effective photonic radar under adverse weather conditions for autonomous vehicles by incorporating a frequency-modulated direct detection scheme," *Frontiers in Physics,* vol. 9, p. 467, 2021.

[122] S. Chaudhary, A. Sharma, and V. Singh, "Optimization of high speed and long haul inter-satellite communication link by incorporating differential phase shift key and orthogonal frequency division multiplexing scheme," *Optik,* vol. 176, pp. 185–190, 2019.

CHAPTER 2

A Study to Discuss the Potential of Mobile Network Optimization Using Machine Learning

Geeta Arora, Jaya Gupta, and Shubham Mishra

2.1 INTRODUCTION

Proactive networks are the cornerstone of future cyber-physical-social systems (CPSS). They offer a unique combination of services: high-speed broadband, ultra-reliable low-latency communication, and massive machine communication, making them ideal for connecting everything in the future of cyber physical social system landscape. 5G and its successors must achieve a staggering 600× to 2,500× increase in capacity, sub-1 mm latency, and support for a mind-boggling 10,000 or more low-rate devices per cell site [1,2]. While significant improvements boost 5G network quality, they also come with a steep increase in operational costs (OPEX) and optimization complexity, which could be nearly 60 times higher according to a study [3]. This poses a serious challenge, as some mobile operators might end up spending more than they earn by the end of the decade if left unchecked [4]. As a result, there are concerns that the high costs associated with maintaining the ambitious quality of 5G heterogeneous networks could significantly outweigh the benefits.

The exponential demands of 5G threaten to be outpaced by soaring operational costs (OPEX) and optimization complexity. However, proactive network optimization offers a transformative solution and

DOI: 10.1201/9781003472506-2

emphasizes the potential for a major shift in how performance and cost are traditionally balanced. One successful example within 3GPP (the 3rd Generation Partnership Project) emphasizes the intelligent management of video content delivery and the use of attention information to gauge how interested users are in something – a promising implementation in CPSS that integrates cloud-edge computing techniques [5]. Data-driven proactive optimization bridges the gap between network resources and predicted user behavior. This allows for anticipating and preventing negative experiences like signal outages or congestion before they occur. For instance, 3GPP Release 15 incorporates user mobility patterns into the Access and Mobility Management Function, enabling prioritized radio resource management [6].

Proactive network optimization in 5G promises a revolution, but challenges remain. Current efforts struggle to gather high-resolution consumer data while accounting for uncertainty across network slices. While online information offers a potential solution, its accuracy and reliability are limited by platform bias and sparsity. The chapter explores the potential of dealing big data analytics of online information with machine learning for proactive network optimization. The study aims to identify demonstrably useful applications and define the boundaries of this approach's effectiveness. The shift from passive/reactive to proactive optimization necessitates the introduction of entirely new network elements, unseen in previous generations (1G–4G) or current 5G standards.

2.1.1 From Reactive to Proactive Network Optimization

The study focuses on radio resource management (RRM) and network deployment, encompassing key optimization functions like scheduling, caching, and interference management. Historically, RRM relied heavily on engineering expertise, with manual configuration dominating the late 1990s. The approach then shifted toward reactive numerical optimization post-2010, incorporating expert-driven functions with parameter inference. To understand the cutting-edge world of proactive optimization, let's start with a quick look back at how optimization has evolved throughout history.

2.1.1.1 Proactive Network Performance Management

Earlier, 2G networks relied heavily on radio engineers who monitored statistics and manually adjusted settings to improve performance metrics (key performance indicators, KPIs). Their decisions were based on field knowledge and experience gained from activities like drive testing [7]. However, this manual approach could take hours to diagnose and fix problems, leading to lengthy network recovery times. 3G ushered in the era of automation to reduce human intervention. Researchers and operators explored solutions like 3GPP's minimization of drive testing (MDT) [8] and Markov decision tree-based optimization for traffic offloading in WCDMA networks [9]. These advancements aimed to reduce reliance on engineers but still required frequent configuration adjustments and weren't personalized to individual users. Optimizations were typically targeted toward broader service areas or service types, like maximizing speed in a city or ensuring fairness across users. Additionally, even these automated schedulers still exhibited significant delays in network recovery, taking over an hour in some cases [10].

2.1.1.2 The Evolving Landscape of Self-Optimization

The era of 4G saw a significant shift with the introduction of self-organizing networks (SON) in 3GPP Release 8 [11]. SON aimed to automate network optimization, enabling cell sites to manage their own coverage, capacity, energy usage, and load balancing [12,13,14,15]. Techniques like reinforcement learning and fuzzy controllers were employed [12,15]. However, a key challenge emerged: the data used by these machine learning approaches often lacked crucial dimensions. Limited information like channel estimates and quality of service reports (QoS) restricted personalization and proactive optimization capabilities [14]. As a result, advanced SON engines typically took over 10 minutes to react to network issues [11,14].

2.1.1.3 Challenges and Solutions for Data Usage Gaps

Recent researches in RRM and network optimization rely heavily on mobile network data (subscriber and cell-level) as the sole source [16,17,18,19]. However, this approach has limitations. While general user behavior is crucial, simplifying it into mathematical models and statistical assumptions offer valuable insights, they may struggle to encompass the full complexity of real-world user needs. Most of the techniques resort to simulating a single user-side measurement, neglecting the rich tapestry of user behavior [3,18,20]. The biggest hurdle lies in gathering and combining multifaceted data from diverse sources, particularly when structured and unstructured formats coexist. This lack of a comprehensive consumer picture creates a "cold start" problem, hindering algorithms that need substantial data for initial optimization.

2.1.2 Unlocking Proactive Optimization with Essential Tools and Technologies

Mobile Data Applications Impacts of 3GPP's report highlights a critical factor for improved network optimization: understanding user behavior and traffic patterns through application data [21]. To achieve this, optimization needs to consider the "context" of users and applications – essentially, the status information of real-world users (UEs) and software applications [22,23]. This context encompasses user location, behavior, and characteristics of network traffic, all crucial for a user-centric network. Techniques for gathering this context are referred to as "context-aware" or "context-awareness" approaches [22,23]. Leveraging cloud-fog-edge computing is essential for processing this context data effectively. To optimize effectively, the algorithm leverages a context-aware module that seamlessly harvests and interprets data from a wide range of sources and personal devices encompassing online data [11]. This rich context then empowers the algorithm to intelligently reallocate communication resources, optimizing network performance for individual users and applications. Some of the essential components are described below.

2.1.2.1 Online Data

Social media platforms are a treasure trove of user-generated data – a rich tapestry of personal information intertwined with location, public updates, and other details [24]. This data offers a direct window into user intent, making it an ideal source for transforming network optimization into an endeavor which is user-centric and proactive. As highlighted in the research, CPSS aim to deliver high-quality services as well as proactive and personalized experiences [25,26]. Here's where big data analysis of online information plays an irreplaceable role. By mining social network data, we can gain valuable insights into user behavior and preferences, enabling networks to anticipate and cater to individual needs. Within the vast realm of online data, this survey delves into three key sources with immense potential for network optimization: social networks such as Instagram, Twitter, and Facebook; reviews and comments of the user (Amazon reviews); and like YouTube some multimedia hosting sites. These platforms offer a rich tapestry of user-generated content, brimming with hidden insights about user behavior and preferences. However, it's important to acknowledge that each data source has its own strengths and weaknesses (merits and pitfalls) that need careful consideration.

2.1.2.2 Social Network Data

The rise of social media has transformed users from passive consumers to active creators and sharers of information. This rich ecosystem provides not only a wealth of public and personal data but also real-time insights for anticipating future network demands [11]. Social network data comes in four primary formats: photos/videos, timeline, text, and geolocation. Text and multimedia content offer a vast amount of information about user preferences, but it can be noisy and unstructured. On the other hand, timeline and geolocation data are well-structured but provide a more limited view of user intent. Social networks also capture social connections, which are invaluable for optimizing device-to-device (D2D) communication. Studies like [27,28] have explored forecasting social ties to enable D2D caching. This involves identifying influential users for content delivery [27] and facilitating content sharing within friend circles [28].

2.1.3 A Data-Driven Framework for Proactive Improvement

Complex computations in our framework rely on a distributed cloud-edge computing architecture, similar to the ones proposed in [25,29,30]. These frameworks partition data processing tasks between the cloud and the edge based on data size and processing needs. The cloud plane, with its superior processing power and storage capacity, handles large-scale, long-term big data analytics (cloud computing) – the "global big data" mentioned earlier. Meanwhile, the edge plane, positioned closer to network users and devices, tackles smaller-scale, real-time data processing tasks (edge computing) – the "local data" processing. Our work adopts a similar approach, utilizing a remote cloud (data-driven proactive optimization module) and an edge cloud, analogous to the cloud and edge planes, respectively. Following a similar approach, our system leverages a remote cloud for data-driven proactive optimization and an edge cloud, mirroring the structure of cloud and edge computing architectures. The remote cloud is positioned between the gateway and the core network, similar to how caching modules are placed near base stations in the core-to-edge network [31]. This distributed approach benefits analysis, learning, and data storage, but it hinges on maintaining acceptably low latency for real-time optimization. The function of the framework is illustrated in the following parts.

2.1.3.1 Data

This crucial step lays the groundwork for effective analysis. It involves data collection, cleaning, and storage. While we focus on these core stages, detailed works further break down the process into data organization, representation, cleaning, reduction, and integration – all well-established practices within CPSS, thanks to the capabilities of edge computing. Data collection involves gathering information relevant to our interests from various offline and online sources, such as government databases, business records, and sensor networks. The data may exhibit variations in completeness and detail across different regions like industrial and urban areas [32]. For internet data, providers often offer application programming interfaces (APIs) that allow authorized access to public datasets. Raw data can be noisy and contain redundancies. A cleaning process (e.g., sorting, filtering) is essential to organize it systematically before storing it in edge or cloud data centers for further analysis. Tensor-based methods, which focus on identifying key features, are effective for big data analysis [33]. These methods can handle both structured and unstructured data through techniques like extensible order tensors. Studies like [34,35] showcase how tensors can be leveraged in cloud-edge computing environments. Big streaming data poses another challenge for real-time processing. This technique is valuable for handling the continuous flow of data in real-time optimization scenarios.

2.1.3.2 Predict Consumer Behaviors

Network traffic is a dynamic beast, constantly evolving with user behavior. This step tackles this challenge by building models to forecast these demand fluctuations. These models leverage online data as input, aiming to predict the probability of different traffic demand levels across diverse user contexts and network slices. By describing the "posterior distribution" of these predictions – essentially the range of potential outcomes – can generate a spatial-temporal map of consumer demand. This map serves as a powerful tool for forecasting key network performance indicators (KPIs).

2.1.3.2 Impact on Network KPIs

Correlation models act as a bridge, translating predicted user behaviors into network KPIs, such as network traffic. Polynomial regression and statistical analysis (e.g., Pearson correlation) are two popular approaches for building these models. These models take the probability of different user behaviors (as predicted in the previous step) as input and generate the probability of various KPI outcomes as output. For instance, the model might predict the probability of a "high-load" scenario occurring in the network.

2.1.3.3 Proactive Network Optimization

The magic of proactive optimization lies in using the predicted network KPIs (from the correlation models) to guide network configuration. The optimization algorithm needs to pre-configure network parameters in anticipation of changing conditions to achieve specific performance targets. For instance, a study in [36] proposes a framework for proactive and energy-efficient network optimization. Their solution leverages an intelligent platform that sits between the core network and the pool of baseband units (BBUs). The platform optimizes the network by analyzing user behavior and traffic patterns. It's important to note that both the remote cloud and the edge plane can contribute to these analytics and learnings:

- **Cloud Computing:** Analyzes large-scale datasets to identify general trends in user behavior, such as city-wide traffic patterns during rush hour. This enables "macroscopic" optimization strategies.
- **Edge Computing:** Processes personalized contexts using data centers, tensors, and data management and analysis techniques based on edge. This personalized element is crucial for user-centric optimization.

2.2 A MULTIFACETED APPROACH: DATA-DRIVEN STRATEGIES FOR NETWORK OPTIMIZATION

Data-driven approaches are revolutionizing network optimization in 5G. Research efforts focus on several key areas:

- **Big Data Analytics:** Studies like [37] explore leveraging big data for 5G network optimization. The authors propose a framework for integrating operator data and showcase optimization use cases.

- **Self-Organizing Network (SON) Architectures:** Pioneering surveys like [38] review cellular SON for beginners, providing an in-depth look at projects, features, standards, and design guidelines.
- **Machine Learning Applications:** Machine learning is a hot topic in SON. Klaine et al. [39] review its applications in this domain, while Chen et al. [40] offer a comprehensive tutorial on using artificial neural networks (ANNs) for wireless communication tasks like proactive caching. This study explores various ANN types, including deep neural networks, recurrent, spiking, and feed-forward.
- **Big Data for Mobility Prediction:** Building on existing big data mobility research [41], this study explores methods for geolocation prediction by analyzing user location patterns and mobility trends. They also compare the suitability of data sources like GPS, GSM, and Wi-Fi records for data analysis and model building in this context.

2.2.1 Essential Techniques for Data Analysis

These methods can be categorized into two broad areas: data collection and data analysis.

2.2.1.1 Data Collection

To understand and forecast social behavior, we delve into three key categories of online data:

1. **Social Networks (e.g., Twitter, Facebook):** These platforms offer a treasure trove of user interactions and public information. Geo-tagged social network data is especially valuable because it reveals users' geographical context, providing a deeper understanding of their activities.
2. **Content Sharing Platforms (e.g., YouTube, Instagram):** These platforms offer insights into user preferences and content consumption habits through the multimedia content they share. By analyzing the types of videos or photos users engage with, we can glean valuable clues about their interests and behaviors.
3. **User Reviews and Ratings (e.g., e-Commerce Reviews, Forum Discussions):** These platforms provide a window into user opinions and sentiment. Analyzing reviews and discussions allows us to understand user experiences and preferences related to specific products or topics.

Traditionally, APIs have been the primary method for collecting this data. APIs act as authorized gateways to specific datasets offered by online platforms. By leveraging APIs, we can gather the information necessary for modeling user activities and their associated metadata – a crucial step for social behavior forecasting.

There are three primary methods for collecting online social network data, each with its own advantages and drawbacks:

- **Application Programming Interfaces (APIs):** This widely used method offers an automated and economical way to collect data directly from service providers. APIs like Search API and Stream API fall under this category. However, limitations include potential inefficiencies, difficulty in acquiring historical data, and data collection limits imposed by service providers.
- **Building Own Datasets:** This approach offers greater control and potentially more complete datasets. Researchers can collect data from volunteers who agree to participate (e.g., study by Xie et al. [42] with 19,484 users) or purchase data from service providers. Incentives like rewards for contribution [43,44] can boost participation but raise costs and require ethical approval.
- **Open Datasets:** Leveraging publicly available datasets from organizations like Kaggle [45] or the European Union Open Data can significantly reduce collection costs. Examples include the Italia Telecom operator dataset for cellular traffic forecasting [46] and datasets from public research projects like DAWN4IoE (cellular traffic data and Twitter density) [47,48,49]. However, researchers need to choose appropriate datasets and pre-process the raw data for further analysis.

2.2.1.2 Data Analysis

The recent rise of scalable machine learning empowers computer systems to learn user behavior from vast online datasets. In the study, there is a discussion of the most commonly used machine learning methods regarding proactive optimization requirements in Figure 2.1 and the categories where these methods have been used. Algorithmic complexity helps us estimate the number of computations required to achieve the desired outcome.

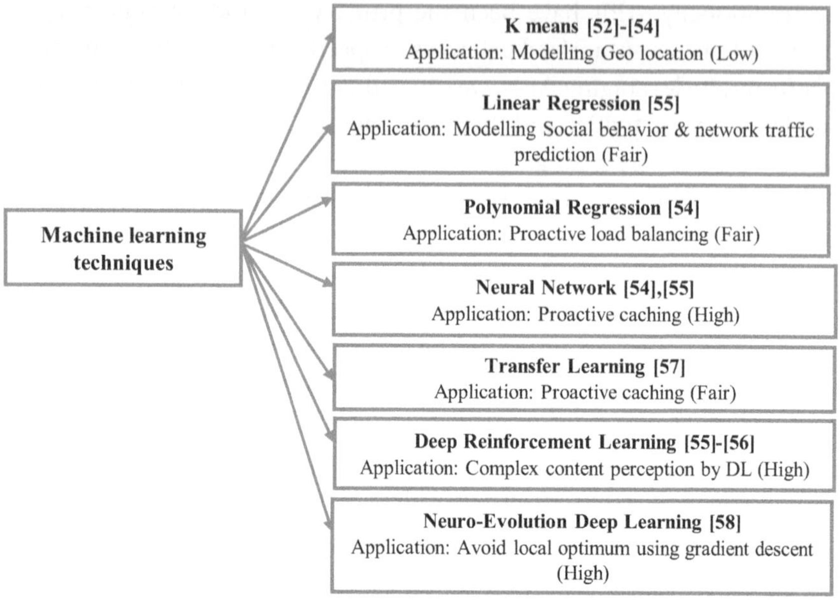

FIGURE 2.1 Types of machine learning, their applications and level of complexity for desired outcomes.

2.2.2 Enhancing Data Utility while Preserving Privacy

Privacy is a paramount concern in data analysis. The challenge lies in maximizing the usefulness (utility) of data while guaranteeing availability, integrity, and confidentiality [50]. Ensuring user data security requires a collaborative effort. Network operators and data providers must prioritize both data encryption and strict access controls to prevent unauthorized data activities. Finding the right balance between data utility and privacy requires a three-pronged approach:

1. **Transparency and User Consent:** Many users are currently unaware of personal data collection, leading to anxieties about potential misuse. Providing clear notices and obtaining informed consent can alleviate these concerns. For instance, the internet of things (IoT) can notify users about data privacy practices (e.g., "right to know") as mandated by regulations like the EU General Data Protection Regulation (GDPR) [51]. However, this approach relies on user trust in data storage and usage practices.

2. **Data Security and Encryption:** Stored data must be protected for both privacy and authenticity. Encryption schemes, such as authenticated encryption [52], can transform data into ciphertext using symmetric-key mechanisms to meet these requirements.
3. **Data Protection and Attack Mitigation:** Data protection goes beyond encryption. Techniques like classifying users based on radio channel information can help identify potentially malicious actors [53]. Research in Hamedani et al. [54] explores using recurrent neural networks (RNNs) to detect various attack variations, while Atat et al. [55] provide a broader survey of security considerations in cyber-physical systems.

2.2.3 Modeling Social Behavior

Understanding user behavior and attitudes is crucial for optimizing communication services. People exhibit a wide range of behaviors and opinions toward different communication methods (e.g., video, audio). By analyzing these behaviors, we can choose the most suitable service type for each user described in the following forms.

2.2.3.1 Content Popularity

Content that users frequently access generates a significant portion of network traffic. Understanding content popularity – essentially, what content is most liked, accessed, and shared – is crucial for optimizing network performance.

There are two main goals for modeling content popularity:

1. **Predicting Popularity:** Forecasting which content will become popular allows for proactive measures to be taken.
2. **Estimating Popularity Levels:** Accurately gauging the level of popularity helps in making informed decisions about transmission, content placement, and storage within proactive caching systems.

Traditional statistical models (like Zipf distribution) for popularity lack the flexibility of machine learning approaches. Machine learning, leveraging online data, offers a promising path forward for building more accurate popularity models. For instance, regression methods can be applied to analyze the popularity of published YouTube videos, enabling proactive caching decisions for new videos. In the context of caching, even small errors can be costly due to high backhaul costs (error tolerance <0.5% as referenced in Chen et al. [56]). Therefore, it's essential to continuously update these models to minimize overhead and optimize resource allocation [57].

2.2.3.2 User Sentiment

User sentiment reflects a user's perception of specific content or services (positive or negative) and can influence their subsequent actions (e.g., continued use, complaints). Understanding user sentiment is valuable for several reasons:

- **Improving Quality of Service (QoS):** By aligning services with user preferences, we can enhance the overall user experience (QoS).
- **Proactive Content Caching:** Identifying preferred content allows for proactive caching, making it readily available to users and reducing network load.

In network optimization, user sentiment can reveal areas with poor performance and content preferences. Social network analysis provides insights into user sentiment, as users often express their experiences (positive or negative) on these platforms. Techniques like sentiment classification can be applied to categorize user complaints on social media according to service type. For instance, a study achieved an F1 score of over 0.6 for blackspot detection in network coverage using Twitter data (as a source for spatial user sentiment analysis) – a case study of London Bridge is presented. Similarly, user requests for specific content reflect their preferences. Studies like [56] explore caching user-preferred content at base stations along their anticipated trajectories.

2.2.3.3 User Relationship

Online platforms like Facebook and Twitter capture the relationships between users through connections like friendships and subscriptions. These relationships can be modeled as graphs, with different types including friendship graphs, interaction graphs, latent graphs, and following graphs. Analyzing these social connections offers valuable insights beyond understanding a user's social role. They also provide a window into information dissemination patterns, which can be harnessed to optimize content delivery [58]. Studies like [58,59] suggest that the strength of social ties influences how information spreads within social-physical networks. In the context of network optimization, leveraging social connections can significantly reduce pressure on the core network. One approach involves strategically sharing cached content within friend circles, identified by analyzing social ties [27]. For instance, a study by Zhu et al. [28] explores using Facebook friendships for cooperative caching.

2.2.3 Using Cellular and Online Information Process of Data-Driven Proactive Network Optimization

The power of online data analysis lies in its ability to reveal hidden patterns and predict user behavior. These insights can be harnessed to forecast cellular traffic patterns, paving the way for a paradigm shift from reactive to proactive network optimization. This section explores how to translate these analytical capabilities into practical implementation strategies for achieving proactive optimization [60].

2.2.3.1 Cellular Traffic Prediction

Accurate network traffic prediction is crucial for proactive network optimization, particularly in proactive load balancing. Traditional methods rely on one-step prediction models built from historical network records, focusing on core network or cell-level predictions. However, these approaches are limited in their ability to predict traffic at finer granularities. A promising solution involves analyzing high-resolution GPS data from various sources (heterogeneous datasets) and correlating it with cellular traffic patterns [61]. Research in Yang et al. [48] confirms a positive correlation between network traffic, online data volume, and the number of network users. This suggests that online social networks can not only predict flash crowd demands but also offer valuable insights for traffic forecasting and resource allocation decisions by network operators.

Traditional methods for spatial traffic prediction rely on mathematical statistics like Zipf distribution and α-stable distributions. However, these methods lack the flexibility of machine learning approaches. Unlike traditional methods, which use techniques like maximum likelihood estimation (approximating parameters without minimizing gradients), machine learning algorithms like neural networks can leverage parameter optimization techniques like gradient descent. While machine learning offers advantages in prediction accuracy, a current challenge is the lack of inherent uncertainty quantification in the generated predictions. This absence of quantifiable uncertainty makes it difficult to assess potential errors and weigh costs and benefits for future decision-making. Promising solutions to address uncertainty quantification include Gaussian processes and deep Gaussian processes, which can produce predictions as posterior distributions that inherently capture uncertainty.

Random traffic spikes pose a significant challenge for network traffic prediction due to the difficulty of explaining and predicting them without additional information about user behavior. To address this limitation, data analysis is moving beyond cellular network data alone and incorporating heterogeneous data sources like Twitter. Current methods often rely on linear regression to model the positive correlation between cellular traffic and social network activity. While statistical methods can quantify the strength of this correlation, they may not effectively capture the underlying dynamics [62, 63]. A critical challenge lies in predicting these random traffic spikes, which often stem from unforeseen events like protests or non-periodic activities. Anomaly detection techniques offer a promising solution to distinguish between regular patterns and anomalous spikes. These techniques often combine clustering algorithms (e.g., K-means) to differentiate traffic patterns and regression models (e.g., neural networks) to predict traffic volume. However, traditional gradient descent optimization in neural networks can lead to locally optimal solutions, potentially hindering prediction accuracy. Neuro-evolutionary deep learning addresses this challenge by incorporating evolutionary techniques to help neural networks escape local optima and achieve better long-term performance with fine-tuned weights.

2.2.4 Proactive Load Balancing

Load balancing is essential for managing network traffic imbalances caused by uneven user demand distribution [40]. It aims to optimize network performance by strategically transferring mobile users (UEs) from congested cells to nearby idle cells. This is achieved through techniques like adjusting handover offset values, thresholds, and the number of handovers themselves [13].

SON algorithms were developed to automate these adjustments, often using fuzzy logic controllers for handover margin tuning [20]. However, these controller-based methods face limitations:

- **Reactive to Traffic Spikes:** They react to sudden traffic spikes, leading to slow response times and limited adaptability to rapidly changing network loads [11].
- **Potential for Oscillations:** In some cases, controller-based methods can introduce oscillations, causing target cells to become overloaded again.

Proactive load balancing approaches leverage machine learning to overcome these challenges. Machine learning algorithms can analyze network data, predict future traffic patterns [64], and proactively adjust key parameters like the call blocking ratio (CBR) to optimize network performance and provides information about the following contents:

(i) Balance load with predictive traffic balance load with call block
(ii) Ratio or UE locations
(iii) Macro cell offloading
(iv) Small cell switch-off
(v) Data-driven user association
(vi) Proactive load balancing based on online information

2.2.5 Proactive Caching

Mobile edge caching algorithms address network traffic demands by strategically storing relevant data on nearby base stations (BSs) based on predicted user needs. This approach significantly improves network performance by enhancing both throughput (data transfer rate) and latency (response time). To maximize the effectiveness of mobile edge caching, proactive consideration of several factors is crucial:

- **Content Selection:** Identifying the most popular or frequently requested content to prioritize for caching.
- **Content Placement:** Deciding which BSs to store specific content on, considering factors like user location and mobility patterns.

- **Content Delivery:** Optimizing the delivery of cached content to users, minimizing delays and network congestion.
- **Storage Utilization:** Efficiently managing cache storage capacity at BSs to accommodate the most relevant content.

By proactively addressing these elements, mobile edge caching algorithms can significantly improve user experience by ensuring content is readily available at the network edge, closer to users and address the following features:

(i) Content placement
(ii) Content delivery
(iii) Storage utilization
(iv) Dynamic caching with mobility prediction
(v) Cold start problem
(vi) Proactive caching based on online information

2.3 CONCLUSION

The study has explored the cutting-edge advancements in online data analytics that hold immense potential for transforming 5G network optimization from a reactive to a proactive approach. The ever-growing volume of online data offers a rich tapestry of information, encompassing individual user contexts (e.g., personal trajectories, preferences, and relationships) and broader social contexts (e.g., popular regions, content popularity, and network traffic patterns). Leveraging machine learning and environmental data, we can predict traffic patterns for different user groups and locations. These forecasts empower proactive network optimization, allowing us to pre-allocate resources to meet diverse needs and adapt to changing network dynamics. This ultimately results in a superior user experience.

REFERENCES

1. J. Yao, T. Han, and N. Ansari, "On mobile edge caching," *IEEE Commun. Surveys Tuts.*, vol. 21, pp. 2525–2553, 3rd Quart., 2019
2. J. G. Andrews, S. Buzzi, W. Choi, S. V. Hanly, A. Lozano, A. C. K. Soong, and J. C. Zhang, "What will 5G be?" *IEEE J. Sel. Areas Commun.*, vol. 32, no. 6, pp. 1065–1082, Jun. 2014.

3. A. Imran and A. Zoha, "Challenges in 5G: How to empower SON wzzzzith big data for enabling 5G," *IEEE Netw.*, vol. 28, no. 6, pp. 27–33, Nov. 2014.
4. Juniper Research Finds. (Nov. 2014). *Mobile Operator Costs Pass$800bn Annually as Margins Become Squeezed.* [Online]. www.juniperresearch.com/press/mobile-operator-costs-pass-800bn-annually-as-margins-become-squeezed-juniper-research-finds/#:~:text=Hampshire%2C%20UK%20%E2%80%93%2018th%20November%202014%3A%20A%20new,the%20end%20of%20the%20decade%20without%20remedial%20action.
5. Aware Service Delivery in RAN for LTE; (Release 14), document 3GPP TR 36.933 V14.0.0, 3rd Generation Partnership Project, 2017.
6. Technical Specification Group; Services and System Aspects; System Architecture for the 5G System; Stage 2 (Release 15), document 3GPP TS 23.501 V0.0.0, 3rd Generation Partnership Project, 2017.
7. A. R. Mishra, *Fundamentals of Cellular Network Planning and Optimisation 2G/2.5G/3G... Evolution to 4G.* Hoboken, NJ, USA: Wiley, 2007.
8. Technical Specification Group TSG RAN Minimization of Drive Tests for E-UTRAN and UTRAN; Overall Description; Stage 2 (Release 10), document 3GPP TS 37.cde V0.3.0, 3rd Generation Partnership Project, 2010.
9. R. Joyce and L. Zhang, "Self-organising network techniques to maximise traffic of road onto a 3G/WCDMA small cell network using MDT UE measurement reports," in *Proc. IEEE Global Commun. Conf.*, Dec. 2014, pp. 2212–2217.
10. D. Fagen, P. A. Vicharelli, and J. Weitzen, "Automated wireless coverage optimization with controlled overlap," *IEEE Trans. Veh. Technol.*, vol. 57, no. 4, pp. 2395–2403, Jul. 2008.
11. A. Aguilar-Garcia, S. Fortes, A. Fernandez Duran, and R. Barco, "Context-aware self-optimization: Evolution based on the use case of load balancing in small-cell networks," *IEEE Veh. Technol. Mag.*, vol. 11, no. 1, pp. 86–95, Mar. 2016.
12. R. Razavi, S. Klein, and H. Claussen, "Self-optimization of capacity and coverage in LTE networks using a fuzzy reinforcement learning approach," in *Proc. 21st Annu. IEEE Int. Symp. Pers., Indoor Mobile Radio Commun.*, Sep. 2010, pp. 1865–1870.

13. I. Viering, M. Dottling, and A. Lobinger, "A mathematical perspective of self-optimizing wireless networks," in *Proc. IEEE Int. Conf. Commun.*, Jun. 2009, pp. 1–6.
14. A. Lobinger, S. Stefanski, T. Jansen, and I. Balan, "Coordinating handover parameter optimization and load balancing in LTE self-optimizing networks," in *Proc. IEEE 73rd Veh. Technol. Conf. (VTC Spring)*, May 2011, pp. 1–5.
15. W. Guo and T. O'Farrell, "Dynamic cell expansion with self-organizing cooperation," *IEEE J. Sel. Areas Commun.*, vol. 31, no. 5, pp. 851–860, May 2013.
16. 3GPP Work Items on Self-Organizing Networks, document v0.1.3, 2014.
17. S. Bassoy, M. Jaber, M. A. Imran, and P. Xiao, "Load aware self-organising user-centric dynamic CoMP clustering for 5G networks," *IEEE Access*, vol. 4, pp. 2895–2906, 2016.
18. B. Keshava Murthy and M. Ashraf, "Conceptual design of pro-active SONs based on the big data framework for 5G cellular networks: A novel machine learning perspective facilitating a shift in the SON paradigm," in *Proc. Int. Conf. Syst. Modeling Adv. Res. Trends (SMART)*, 2016, pp. 298–304.
19. W. Zuo, H. Xie, Y. Lin, H. Hu, and Z. Cai, "A load balancing optimization algorithm for context-aware wireless sensor networks based on fuzzy neural networks," *Amer. J. Neural Netw. Appl.*, vol. 2, no. 2, pp. 6–16, 2016.
20. L.-C. Wang, S.-H. Cheng, and A.-H. Tsai, "Bi-SON: Big-data self-organizing network for energy efficient ultra-dense small cells," in *Proc. IEEE 84th Veh. Technol. Conf. (VTC-Fall)*, Sep. 2016, pp. 1–5.
21. Technical Specification Group Services and System Aspects; Study on Non-MTC Mobile Data Applications Impacts (Release 11), document 3GPP TR 22.801 V2.0.0, 3rd Generation Partnership Project, 2011.
22. J. Wu, I. Bisio, C. Gniady, E. Hossain, M. Valla, and H. Li, "Context aware networking and communications: Part 1 [Guest Editorial]," *IEEE Commun. Mag.*, vol. 52, no. 6, pp. 14–15, Jun. 2014.
23. J. Wu, I. Bisio, C. Gniady, E. Hossain, M. Valla, and H. Li, "Context aware networking and communications: Part 2 [Guest Editorial]," *IEEE Commun. Mag.*, vol. 52, no. 8, pp. 64–65, Aug. 2014.

24. J. Krumm, N. Davies, and C. Narayanaswami, "User-generated content," *IEEE Pervasive Comput.*, vol. 7, no. 4, pp. 10–11, Oct. 2008.
25. X. Wang, L. T. Yang, X. Xie, J. Jin, and M. J. Deen, "A cloud-edge computing framework for cyber-physical-social services," IEEE Commun. Mag., vol. 55, no. 11, pp. 80–85, Nov. 2017.
26. X. Wang, L. T. Yang, H. Li, M. Lin, J. Han, and B. O. Apduhan, "NQA: Anested anti-collision algorithm for RFID systems," *ACM Trans. Embedded Comput. Syst.*, vol. 18, no. 4, pp. 1–21, Jul. 2019.
27. E. Bastug, M. Bennis, and M. Debbah, "Living on the edge: The role of proactive caching in 5G wireless networks," *IEEE Commun. Mag.*, vol. 52, no. 8, pp. 82–89, Aug. 2014.
28. K. Zhu, W. Zhi, L. Zhang, X. Chen, and X. Fu, "Social-aware incentivized caching for D2D communications," *IEEE Access*, vol. 4, pp. 7585–7593, 2016.
29. X. Wang, L. T. Yang, J. Feng, X. Chen, and M. J. Deen, "A tensor-based big service framework for enhanced living environments," *IEEE Cloud Comput.*, vol. 3, no. 6, pp. 36–43, Nov. 2016.
30. X. Wang, L. T. Yang, H. Liu, and M. J. Deen, "A big data-as-a-service framework: State-of-the-art and perspectives," *IEEE Trans. Big Data*, vol. 4, no. 3, pp. 325–340, Sep. 2018.
31. D. Liu, B. Chen, C. Yang, and A. F. Molisch, "Caching at the wireless edge: Design aspects, challenges, and future directions," *IEEE Commun. Mag.*, vol. 54, no. 9, pp. 22–28, Sep. 2016.
32. N. Gupta, H. Crosby, D. Purser, S. Javis, and W. Guo, "Twitter usage across industry: A spatiotemporal analysis," in *Proc. IEEE 5th Int. Conf. Big Data Comput. Service Appl. (Big Data Service)*, Mar. 2018, pp. 64–71.
33. X. Wang, L. T. Yang, X. Chen, J.-J. Han, and J. Feng, "A tensor computation and optimization model for cyber-physical-social big data," *IEEE Trans. Sustain. Comput.*, vol. 4, no. 4, pp. 326–339, Oct. 2019.
34. X. Wang, L. T. Yang, L. Kuang, X. Liu, Q. Zhang, and M. J. Deen, "A tensor-based big-data-driven routing recommendation approach for heterogeneous networks," *IEEE Netw.*, vol. 33, no. 1, pp. 64–69, Jan. 2019.
35. L. Kuang, L. T. Yang, X. Wang, P. Wang, and Y. Zhao, "A tensor-based big data model for QoS improvement in software defined networks," *IEEE Netw.*, vol. 30, no. 1, pp. 30–35, Jan. 2016.

36. Y. Li, Y. Zhang, K. Luo, T. Jiang, Z. Li, and W. Peng, "Ultra-dense Het Nets meet big data: Green frameworks, techniques, and approaches," *IEEE Commun. Mag.*, vol. 56, no. 6, pp. 56–63, Jun. 2018.
37. K. Zheng, Z. Yang, K. Zhang, P. Chatzimisios, K. Yang, and W. Xiang, "Big data-driven optimization for mobile networks toward 5G," *IEEE Netw.*, vol. 30, no. 1, pp. 44–51, Jan. 2016.
38. O. G. Aliu, A. Imran, M. A. Imran, and B. Evans, "A survey of self-organisation in future cellular networks," *IEEE Commun. Surveys Tuts.*, vol. 15, no. 1, pp. 336–361, 1st Quart., 2013.
39. P. V. Klaine, M. A. Imran, O. Onireti, and R. D. Souza, "A survey of machine learning techniques applied to self-organizing cellular networks," *IEEE Commun. Surveys Tuts.*, vol. 19, no. 4, pp. 2392–2431, 4th Quart., 2017
40. M. Chen, U. Challita, W. Saad, C. Yin, and M. Debbah, "Artificial neural networks-based machine learning for wireless networks: A tutorial," 2017, arXiv:1710.02913. [Online]. Available: http://arxiv.org/abs/1710.02913
41. G. Xu, S. Gao, M. Daneshmand, C. Wang, and Y. Liu, "A survey for mobility big data analytics for geolocation prediction," *IEEE Wireless Commun.*, vol. 24, no. 1, pp. 111–119, Feb. 2017.
42. R. Xie, Y. Chen, Q. Xie, Y. Xiao, and X. Wang, "We know your preferences in new cities: Mining and modeling the behavior of travellers," *IEEE Commun. Mag.*, vol. 56, no. 11, pp. 2835, Nov. 2018.
43. F. Zeng, R. Wang, and J. Wu, "How mobile contributors will interact with each other in mobile crowdsourcing with word of mouth mode," *IEEE Access*, vol. 7, pp. 14523–14536, 2019.
44. A. Tsakalidis, M. Liakata, T. Damoulas, B. Jellinek, W. Guo, and A. I. Cristea, "*Combining Heterogeneous User Generated Data to Sense Well-Being*," Osaka, Japan: The COLING 2016 Organizing Committee, 2016.
45. Datasets on Kaggle. Accessed: Feb. 19, 2020. [Online]. Available: https://www.kaggle.com/datasets
46. C. Zhang, X. Ouyang, and P. Patras, "ZipNet-GAN: Inferring dengrained mobile traffic patterns via a generative adversarial neural network," in *Proc. 13th Int. Conf. Emerg. Netw. Exp. Technol. (CoNEXT)*, 2017, pp. 363–375.

47. G. Khodabandelou, V. Gauthier, M. Fiore, and M. A. El-Yacoubi, "Estimation of static and dynamic urban populations with mobile network metadata," *IEEE Trans. Mobile Comput.*, vol. 18, no. 9, pp. 2034–2047, Sep. 2019.
48. B. Yang, W. Guo, B. Chen, G. Yang, and J. Zhang, "Estimating mobile traffic demand using Twitter," *IEEE Wireless Commun. Lett.*, vol. 5, no. 4, pp. 380–383, Aug. 2016.
49. B. Yang, W. Guo, B. Chen, G. Yang, and J. Zhang, Data From: Estimating Mobile Traffic Demand Using Twitter. 2016. Accessed: April 19, 2021 [Online]. Available: https://doi.org / 10.5061/dryad.35m1f4q
50. L. M. Kaufman, "Data security in the world of cloud computing," *IEEE Secur. Privacy Mag.*, vol. 7, no. 4, pp. 61–64, Jul. 2009.
51. R. Chow, "The last mile for IoT privacy," *IEEE Secur. Privacy*, vol. 15, no. 6, pp. 73–76, Nov. 2017.
52. D. Maimut and R. Reyhanitabar, "Authenticated encryption: Toward next-generation algorithms," *IEEE Secur. Privacy*, vol. 12, no. 2, pp. 70–72, Mar. 2014.
53. S. Chen, H. Wen, J. Wu, J. Chen, W. Liu, L. Hu, and Y. Chen, "Physical layer channel authentication for 5G via machine learning algorithm," *Wireless Commun. Mobile Comput.*, vol. 2018, pp. 1–10, Oct. 2018.
54. K. Hamedani, L. Liu, R. Atat, J. Wu, and Y. Yi, "Reservoir computing meets smart grids: Attack detection using delayed feedback networks," *IEEE Trans Ind. Informat.*, vol. 14, no. 2, pp. 734–743, Feb. 2018.
55. R. Atat, L. Liu, J. Wu, G. Li, C. Ye, and Y. Yang, "Big data meet cyber-physical systems: A panoramic survey," *IEEE Access*, vol. 6, pp. 73603–73636, 2018.
56. M. Chen, W. Saad, C. Yin, and M. Debbah, "Echo state networks for proactive caching in cloud-based radio access networks with mobile users," *IEEE Trans. Wireless Commun.*, vol. 16, no. 6, pp. 3520–3535, Jun. 2017.
57. S. Ioannidis, A. Chaintreau, and L. Massoulié, "Distributing content updates over a mobile social network," *ACM SIGMOBILE Mobile Comput. Commun. Rev.*, vol. 13, no. 1, p. 44, Jun. 2009.
58. Y. Yi, Z. Zhang, and C. Gan, "The effect of social tie on information diffusion in complex networks," *Phys. A Stat. Mech. Appl.*, vol. 509, pp. 783–794, Nov. 2018.

59. Y. Yi, Z. Zhang, and C. Gan, "The outbreak threshold of information diffusion over social physical networks," *Phys. A Stat. Mech. Appl.*, vol. 526, Jul. 2019, Art. no. 121128.
60. C. Qiu, Y. Zhang, Z. Feng, P. Zhang, and S. Cui, "Spatio-temporal wireless traffic prediction with recurrent neural network," *IEEE Wireless Commun. Lett.*, vol. 7, no. 4, pp. 554–557, Aug. 2018.
61. B. Yang, W. Guo, B. Chen, G. Yang, and J. Zhang, "Estimating mobile traffic demand using Twitter," *IEEE Wireless Commun. Lett.*, vol. 5, no. 4, pp. 380–383, Aug. 2016.
62. A. Sathiaseelan, M. S. Seddiki, S. Stoyanov, and D. Trossen, "Social SDN: Online social networks integration in wireless network provisioning," *Proc. ACM Conf. (SIGCOMM), New York, NY, USA*, vol. 44, no. 4, pp. 375–376, 2014.
63. B. Yang, W. Guo, B. Chen, G. Yang, and J. Zhang, "Estimating mobile traffic demand using Twitter," *IEEE Wireless Commun. Lett.*, vol. 5, no. 4, pp. 380–383, Aug. 2016.
64. A. Lobinger, S. Stefanski, T. Jansen, and I. Balan, "Load balancing in downlink LTE self-optimizing networks," in *Proc. IEEE 71st Veh. Technol. Conf.*, May 2010, pp. 1–5.

CHAPTER 3

AI in Mobile Network Management and Performance Enhancement

Avani Vyas and Shivani Sharma

3.1 INTRODUCTION

Mobile networks are a telecommunication network of mobile devices such as smartphones, tablets, and other wireless-enabled devices that can connect to a network. Mobile networks are also termed as cellular networks [1]. A geographical region is divided into cells. Each cell contains a base station which is a physical structure equipped with antennas and transceivers. A base station provides connectivity to mobile devices present in its cell with the rest of the cellular network. The network size and services provided by mobile devices are growing with the growing wireless communication technology. The various communication services supported by mobile networks include voice calls, text messaging (SMS), multimedia messaging (MMS), and data services like internet access. Mobility of the user devices challenges the quality of service (QoS) provided by communication networks. This chapter discusses the key aspects involved in mobile network management and the use of AI in improving users' mobile network experience.

DOI: 10.1201/9781003472506-3

3.1.1 Mobile Network Management

The key aspects of mobile network management include:

1. **Monitoring Network:** It involves maintaining network performance parameters and fault tolerance. The network performance matrices of prime importance are latency, packet loss, and bandwidth utilization. Whereas fault tolerance involves detecting and identifying network faults to minimize downtime and service disruptions.
2. **Traffic Management:** Traffic management handles load scheduling in the mobile network in order to optimize the flow of data to prevent congestion and enhance resource utilization.
3. **Quality-of-Service (QoS) Management:** The QoS of a mobile network depends on traffic management and network monitoring. A well-distributed network traffic load with optimal resource utilization is expected to improve the network performance parameters such as latency, packet loss, and bandwidth utilization.
4. **Security Management Services:** In the wide connectivity of a network, it is crucial to implement a security mechanism. Security management in mobile networks involves preventing unauthorized access and controlling access to the network and its resources.

3.1.2 Scope of Artificial Intelligence in Mobile Network Management

Artificial intelligence (AI) techniques have made a profound contribution to mobile network management by enhancing the network lifetime, efficiency, and security requirements [2,3]. AI techniques like predictions, machine learning (ML), deep learning, expert systems, and reinforcement learning can be applied to mobile network management in the following ways:

1. **Predictive Maintenance:** AI-based data analysis tools can be used for predicting network failures and maintenance of networking devices along with real-time monitoring. Predictions help in implementing proactive maintenance strategies for addressing issues that may lead to service disruptions.
2. **Dynamic Resource Allocation:** The ML models can be implemented in mobile networks for analyzing traffic patterns and dynamic

allocation of network resources such as bandwidth to meet the changing demands of users. Overall, ML algorithms can be designed for intelligent traffic management in mobile networks.

3. **Designing Self-optimizing Networks:** AI algorithms are used to automatically optimize network parameters such as antenna configurations and power levels for improving overall user experience.
4. **Improving Network Security Aspects:** AI/ML algorithms are used for detecting unusual traffic patterns and anomaly detection and identifying unauthorized access in the network. Similarly, AI algorithms are used for generating automated responses in case of identification of unauthorized behaviors.
5. **AI for Green Communication:** Predictive AI is very useful in designing green communication networks. Various network components can be programmed for smart sleep modes so that they can optimize the use of their battery powers resulting in prolonged network services.
6. **AI for Future Generation Communication:** AI techniques can be leveraged for optimizing beamforming, network slicing, and edge computing.

3.2 TRADITIONAL METHODS OF MOBILE NETWORK MANAGEMENT

This section discusses the non-AI-based traditional methods designed for mobile network management.

3.2.1 Mobile Network Optimization

The objective of mobile network optimization is to enhance network efficiency, maximize resource utilization, reduce congestion, and ultimately improve QoS. The targets are managed by optimizing radio frequency planning, implementing load balancing, and employing algorithms to manage traffic and resources more efficiently. Rodriguez et al. [4] addressed the problem of increasing network management with the increasing number of mobile users in 5G networks. Challenges majorly include reusing the radio spectrum and handling handovers with the increasing network classifications. The authors used a stochastic geometry approach on radio environment maps to improve the association and coverage probability of the network. The Monte Carlo simulation results show improved coverage probability and handover rates compared to other cell-association techniques. Saad et al. [5] also address the same problem of increasing mobile users on the limited availability of cells. The authors investigated trades and tuning of various handover control parameters settings. They concluded that automatic self-optimization is the best for handling such challenges rather than manual handling. Load balancing is a well-explored problem in mobile networks that involves balancing traffic between hot cells (overloaded cells) and cooler cells (under-loaded cells) [6–8]. The traditional methods of load balancing include information-centric networking, cost function-based algorithms [9], and parameter-converging algorithms [10].

3.2.2 Mobile Network Performance Management

The objective of mobile network performance management is to identify the areas of improvement and continuous monitoring of metrics such as data throughput, latency, and network availability. Sherif et al. [11] improved the performance of mobile networks by addressing the issue of selfish nodes in opportunistic mobile networks (OMNs). OMNs involve communication between nodes that may not have continuous connectivity. The communication in such networks depends on opportunistic encounters of mobile nodes which sometimes increases communication latency. Therefore, intermediate or relay nodes are used for relaying the messages of disconnected nodes to increase the data throughout and reduce latency. However, sometimes, the intermediate relays may act as selfish and may not cooperate in relaying messages, affecting the network performance and user experience. Sherif et al. [11] and Jedari et al. [12] used watchdog mechanisms for monitoring and detecting misbehaving nodes in OMNs. Over the years, many such algorithms have been proposed for detecting and handling misbehaving nodes in mobile networks. Kampitaki et al. [13] provide a comprehensive survey of such algorithms.

3.2.3 Mobile Network Resource Allocation

Mobile devices contain limited energy and computation resources. However, modern applications like real-time tracking systems demand high computation and continuous battery support. In order to support resource-hungry mobile applications, a mechanism of task offloading is used in networks. Task offloading is a process of transferring computational tasks from a mobile device to another computing resource, typically a more powerful server. The results of the computation are then sent back to the mobile device from the server [14]. However, transferring computation tasks from mobile devices to servers and back to mobile devices increases communication latency in applications. Radio resource allocation methods are required when multiple mobile devices request task offloading in a cell [15]. The solutions proposed include optimal sub-channel allocation [15], a heuristic performance-based priority approach [14], and multiple parameters that can be combined into non-linear programming problems [16].

In conclusion, the traditional methods of mobile network management include designing multi-objective functions and using optimization algorithms like Watchdog and game theory-based approaches for optimizing the performance of the network. However, these methods are difficult to update with the changing network dimensions and users. The modern networks are time-varying and demand self-adaptive solutions for accommodating modern mobile applications.

3.3 ARTIFICIAL INTELLIGENCE FOR MOBILE NETWORKS

This section discusses in detail the existing uses of AI in mobile network management.

3.3.1 AI-based Mobile Network Optimization

3.3.1.1 Self-healing Networks

Self-healing refers to the ability to detect, diagnose, and mitigate the issue automatically without human intervention. Moysen et al. [17] designed a reinforcement learning-based automated self-healing approach for identifying and diagnosing cell outages in long-term evolution (LTE) networks. The AI-enabled base station automatically adjusts the downlink transmission power levels and finds the optimal antenna tilt value for covering the outage zone. Asghar et al. [18] used a ML algorithm on network performance matrices to continuously detect anomalies in the network. After detection, fuzzy logic is used to automatically determine the compensation actions. Additionally, a heuristic-based continuous optimization tool is designed for making incremental improvements to the network. Reshmi et al. [19] also designed a self-healing framework that uses network performance matrices for detecting anomalies in the network. Riaz et al. [20] discussed that the existing efforts in self-healing networks focus on single base stations and single faults. Moreover, they highlighted that the existing AI-based methods fail to perform if dense user datasets are not available during training. In order to overcome the research gap, Riaz et al. [20] designed a self-healing framework using CNN and image in painting technique which outperforms many ML models on highly sparse user equipment datasets.

3.3.1.2 Energy Efficiency

Hossain et al. [21] forecasted the changing network load and expected energy utilization based on which network slicing decisions are made. They distributed the computation load of ML algorithms among different computation devices available in the network to reduce the energy consumed by different ML algorithms. Peesapati et al. [22] addressed the concept of multi-operator radio access network (RAN) sharing, where multiple mobile network operators collaboratively utilize the same physical infrastructure for their radio access networks. This sharing model is aimed at resource efficiency and cost reduction. AI is employed to make intelligent decisions and adjustments to the network configuration with the goal of reducing energy consumption while maintaining performance.

3.3.1.3 Load Balancing

AI has found great importance in load balancing in mobile networks. Gures et al. [23] provide a detailed survey of such load-balancing techniques. Alsuhli et al. [24] handled the issue of mobility load management in cellular networks. The focus of the research is on dynamically allocating network resources to accommodate the mobility patterns of users. The authors used deep reinforcement learning for dynamically allocating network resources to accommodate the mobility patterns of users. Saad et al. [25] highlighted that the cell density is increasing with the increasing user equipment which demands optimized handovers of mobile devices. The authors in [26] designed a self-optimizing fuzzy logic-based handover scheme for heterogeneous networks. Three parameters, namely the user speed, SINR value, and load of the cell are input into the designed fuzzy system for estimating the handover control parameters.

3.3.2 Intelligent Traffic Management in Mobile Networks

Intelligent traffic management in mobile networks involves the use of advanced technologies and algorithms to optimize the flow of data and resources within the network. The goal is to enhance the overall performance, efficiency, and quality of service in the face of varying network conditions, user demands, and application requirements. Intelligent traffic management encompasses a range of techniques and strategies to dynamically allocate resources, prioritize traffic, and respond to real-time network conditions. Traditional methods of mobile network/internet traffic management exploited queuing theory for handling bottleneck buffer queues in mobile networks [27]. With the introduction of AI, researchers have started exploiting AI methods for handling network traffic. Liu et al. [28] designed a fuzzy-logic-based controller for automatically measuring router queue size. The designed model helps in taking proactive measures based on measured router queue sizes. Fu et al. [29] discussed the services provided with a 5G network along with the potential use of AI for improving network performance. The authors mentioned the potential deployment of distributed ML algorithms at cloud, fog, and edge nodes for 5G networks. Shen et al. [30] discussed the issue of QoS in heterogeneous networks. In heterogeneous networks, different types of networks are combined using connecting devices like gateways. Different networks have different traffic rates and service requirements. Shen et al. [30] performed classification and management at different layers of convolution neural networks for designing intelligent multi-dimensional heterogeneous traffic control.

AI techniques are also used in prediction-based traffic management [31]. Morales et al. [32] used artificial neural networks to predict the network traffic and adapt the topology of virtual networks in order to support the increased traffic demand. Virtual networks provide extensions to mobile networks by adding more computation power to the network. Morales et al. [32] predicted the maximum bit rate between every origin-destination pair on an hourly basis and adapted the topologies of the virtual network to meet the predicted traffic demand. Similarly, AI techniques are used in optical networks. Optical networks provide backhaul connectivity to mobile networks for transporting mobile network traffic between various network elements, including base stations, core networks, and external data centers [33–35]. This ensures efficient and reliable data transport. Leung et al. [36] used a single-hidden-layer neural network for estimating the blocking probability of current network traffic conditions in optical networks. In another contribution by Vinchoff et al. [37], they exploited graph convolution neural network (GCN) and generative adversarial network (GAN) for predicting the high-traffic situations in optical networks.

The study shows that the GCN-GAN model predicts traffic bursts more accurately compared to the traditional LSTM model.

3.3.3 Machine Learning for Mobile Network Planning

Mobile network planning involves designing, optimizing, and configuring the elements of a mobile communication network to ensure efficient and reliable service delivery.

3.3.3.1 Network Coverage Planning

Fauzi et al. [38] highlighted that the deterministic techniques of network planning are failing to meet the increasing demands of mobile networks. They claimed that ML algorithms are useful in estimating the propagation models over the traditional empirical methods. The authors used a random forest ML algorithm for predicting the reference signal received power and designed an online coverage estimator tool. Dai et al. [39] also emphasized using ML algorithms for estimating network coverage over the traditional propagation models as they are complex and time-consuming. They also designed an ML-based network evaluation tool by considering useful parameters such as communication distance, geography, and configuration parameters of the base station. Different ML models are trained on real-world datasets for estimating the received signal strength (RSS). The comparison results on the considered dataset show that SVM achieved better RSS prediction accuracy as compared to the nearest neighbors, decision trees, random forest, and neural network models. Mohammadjafari et al. [40] used ML algorithms to estimate the radio coverage and determine optimal radio transmitter locations for a given geographical location. They investigated that creating and adding new features based on the existing features to the ML models further improves the prediction capabilities of models. Mahboubi et al. [41] used K-means clustering to increase the area covered by a mobile sensor network. Lyu et al. [42] used a supervised learning model for estimating the coverage of mobile networks. They also used AI to generate synthesized datasets and combine them with the real-world dataset. The final dataset is then used for training and testing the supervised learning models. Likewise, over the years, numerous ML techniques have been designed and explored for enhancing cellular networks [43].

3.3.3.2 Backhaul Planning

Backhaul is the part of the network that connects the core or backbone network to the small subnetworks at the network edge, typically the access networks. Backhaul must have sufficient capacity to handle the data traffic generated by the connected access points. With the increasing demand for high-speed data services, the capacity of backhaul links becomes crucial for ensuring a responsive and efficient network. Mahmood et al. [44] used an ML algorithm for predicting future network demands. The prediction helped in dynamically configuring the wireless backhaul network based on the forecasted traffic. This adaptability allows the network to respond in real time to changing demand patterns and ensures that resources are allocated where they are most needed. Abdelmoaty et al. [45] used deep reinforcement learning for designing hierarchical backhaul topology.

3.4 USE CASES AND APPLICATIONS

Several telecommunication industries are shifting toward AI-enabled services. As discussed earlier in the chapter, the main goal is to provide automated self-healing and adaptive services to users for handling seamless handovers and network outages. In a recent survey report by Frost and Sullivan [46], 71% of the surveyed telecommunication companies experience improved customer services and 63% companies optimized their network management using AI. German ministry funded 5.1 million Euros to Ericsson project "AI4Mobile" [47]. "AI4Mobile" provides improved results for the prediction of connection quality for Connected Mobility and Industry 4.0. The existing contribution of AI in supporting mobile networks is not limited to this discussion. Digital twins are emerging as the most powerful AI tool that can be used in many ways to improve the services of telecommunications.

3.4.1 Digital Twins for Mobile Networks

A digital twin is a virtual representation or digital replica of a physical object, system, or process. It is a concept that combines real-world data with virtual modeling and simulation to create a dynamic, digital counterpart of a physical entity. The goal is to enable monitoring, analysis, and optimization of the physical entity throughout its lifecycle. The digital twin provides a facility for testing a process virtually on a real-life platform without actually causing any physical damage or cost. Sun et al. [48] designed a digital-twin network of a mobile network for the fast management of service level agreements (SLA). All the updated network policies are first verified on the digital-twin network for SLA requirements without affecting the physically deployed network services. Figure 3.1 shows the digital-twin network adaptation of a real physical network.

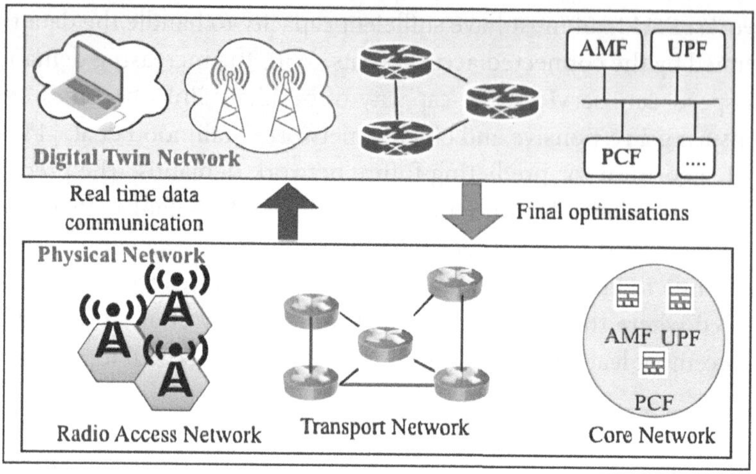

FIGURE 3.1 Digital twins for mobile networks [48].

Deng et al. [49] address that the performance of ML algorithms can be improved further by integrating digital-twin networks into them. The current proposed adaptive ML techniques rely on physical networks for training samples and adaptations which causes delays in data collection, training, and deployment. As a solution, the authors considered the digital twin of the current physical network and used expert knowledge to predict the next state of the network. Then, reinforcement learning is used for optimization. The final optimized solution is deployed on the physical network. Likewise, digital twins have found their potential use in designing and testing security and privacy services in mobile networks [50]. Ahmadi et al. [51] discussed the application of digital twins in 6G networks. They discussed the concept of "networked twins" which refers to a collaborative ecosystem of interconnected digital twins within a network. This includes digital twins of individual devices, components, or even entire systems interacting with each other, sharing data and collectively contributing to the optimization of the entire network.

3.5 EMERGING TECHNOLOGIES

3.5.1 Edge Computing and AI in Mobile Networks

Edge computing involves processing data closer to the source of generation, re-reducing latency, and improving real-time responsiveness. The combination of edge computing and AI facilitates real-time decision-making at the network's edge. This is particularly valuable for applications requiring immediate responses, such as autonomous vehicles and smart cities. AI algorithms like neural networks demand high processing capabilities. The mobile devices are equipped with limited computing resources. The major processing tasks such as periodic training of adaptive AI models are usually performed at resource-rich cloud networks. However, mobile networks experience communication latency due to the exchange of data and training model parameters between mobile devices and the cloud. Therefore, the major work in the area lies in exploiting distributed edge resources for AI mobile networks to improve response speed. Li et al. [52] partitioned the computation of deep neural networks between edge and mobile devices for real-time adaptations with improved user experience. Tran et al. [53] presented the idea of upgrading the mobile network infrastructure by integrating mobile edge computing (MEC) capabilities directly at the base stations for further optimizing the services supported by edge networks. The proposed infrastructure was evaluated for three case studies which include a compute-intensive application, pre-processing of large data before sending it to the servers, and context-aware services. As a result, the designed heterogeneous resource pool of MEC and edge proved its potential application in 5G networks.

3.5.2 AI for Network Security in Mobile Communication

Security threats in mobile networks pose risks to the confidentiality, integrity, and availability of data and services. As mobile networks evolve, new threats emerge, and existing threats may evolve in sophistication. To name a few, the common security threats in mobile networks include network spoofing, unauthorized and authentication attacks, eavesdropping, and denial of service attacks. The conventional methods of cryptography and security find it challenging to provide services in complex time-varying mobile networks. This provides a scope for artificial intelligence-based methods to address time-varying parameters and provide automated and adaptive security mechanisms. Fang et al. [54] explored ML-based authentication methods for 5G networks that exploit physical layer parameters such as carrier frequency offset, channel impulse response, and RSSI values. The pattern recognition algorithms are useful in identifying anomalies in time-series datasets which could lead to security threats in mobile networks. Ali et al. [55] identified anomalies due to radio network attenuation and SDN misconfiguration using a clustering algorithm. Jamming attacks are a common threat in mobile networks resulting in denial of service. Krayani et al. [56] used a dynamic Bayesian network model to detect jamming in orthogonal frequency division multiplexing subcarriers. Though AI methods are useful in improving the services of mobile networks, they are vulnerable to security attacks. Qiu et al. [57] tested a deep neural network model for white-box and black-box attacks and proved that the AI models can be fooled by modifying their parameters or by tampering with the datasets. This demands further actions for addressing vulnerabilities in AI-based security management in mobile networks.

3.6 CONCLUSION

This chapter discusses the use of ML models, predictive analysis, neural network models, and reinforcement learning in enhancing mobile network management services. This chapter provides an overview of the traditional methods of mobile network management followed by a discussion on the latest AI techniques used for designing self-healing mobile networks, automatic load rebalancing, automated network, and backhaul planning. Digital-twin technology has the potential to be transformative for mobile networks, offering substantial positive impacts. With the use of digital twins, tedious real-time tests can be performed with reduced complexities and time.

REFERENCES

[1] M. Sauter, *From GSM to LTE: An Introduction to Mobile Networks and Mobile Broadband*. Hoboken: John Wiley & Sons, 2010.

[2] A. Havolli and M. Fetaji, "Improving radio network planning and design in next-generation mobile networks using AI and ML algorithms," in *2023 12th Mediterranean Conference on Embedded Computing (MECO)*, pp. 1–5, 2023.

[3] G. Luo, Q. Yuan, J. Li, S. Wang, and F. Yang, "Artificial intelligence powered mobile networks: From cognition to decision," *IEEE Network*, vol. 36, no. 3, pp. 136–144, 2022.

[4] A. C. S. Rodriguez, N. Haider, Y. He, and E. Dutkiewicz, "Network optimisation in 5g networks: A radio environment map approach," *IEEE Transactions on Vehicular Technology*, vol. 69, no. 10, pp. 12043–12057, 2020.

[5] W. K. Saad, I. Shayea, B. J. Hamza, H. Mohamad, Y. I. Daradkeh, and W. A. Jabbar, "Handover parameters optimisation techniques in 5g networks," *Sensors*, vol. 21, no. 15, p. 5202, 2021.

[6] A. Sharma, A. Roy, S. Ghosal, R. Chaki, and U. Bhattacharya, "Load balancing in cellular network: A review," in *2012 Third International Conference on Computing, Communication and Networking Technologies (ICCCNT'12)*, pp. 1–5, IEEE, 2012.

[7] T. Semong, T. Maupong, S. Anokye, K. Kehulakae, S. Dimakatso, G. Boipelo, and S. Sarefo, "Intelligent load balancing techniques in software defined networks: A survey," *Electronics*, vol. 9, no. 7, p. 1091, 2020.

[8] M. R. Belgaum, S. Musa, M. M. Alam, and M. M. Su'ud, "A systematic review of load balancing techniques in software-defined networking," *IEEE Access*, vol. 8, pp. 98612–98636, 2020.

[9] Y. Yang, P. Li, X. Chen, and W. Wang, "A high-efficient algorithm of mobile load balancing in LTE system," in *2012 IEEE Vehicular Technology Conference (VTC Fall)*, pp. 1–5, IEEE, 2012.

[10] Q. Ye, B. Rong, Y. Chen, M. Al-Shalash, C. Caramanis, and J. G. Andrews, "User association for load balancing in heterogeneous cellular networks," *IEEE Transactions on Wireless Communications*, vol. 12, no. 6, pp. 2706–2716, 2013.

[11] B. V. Sherif and P. Salini, "Selfish node management in opportunistic mobile networks with improved social based watchdog and dynamic power AODV protocol," *International Journal of Information Technology*, vol. 14, p. 3253–3264, 2022.

[12] B. Jedari, F. Xia, H. Chen, S. K. Das, A. Tolba, and Z. AL-Makhadmeh, "A social-based watchdog system to detect selfish nodes in opportunistic mobile networks," *Future Generation Computer Systems*, vol. 92, pp. 777–788, 2019.

[13] D. G. Kampitaki and A. A. Economides, "Selfishness in mobile ad-hoc networks: A literature review on detection techniques and prevention mechanisms," *IEEE Access*, vol. 11, pp. 86895–86909, 2023.

[14] T.-Y. Kan, Y. Chiang, and H.-Y. Wei, "Task offloading and resource allocation in mobile-edge computing system," in *2018 27th Wireless and Optical Communication Conference (WOCC)*, pp. 1–4, IEEE, 2018.

[15] Y. Cao, T. Jiang, and C. Wang, "Optimal radio resource allocation for mobile task offloading in cellular networks," *IEEE Network*, vol. 28, no. 5, pp. 68–73, 2014.

[16] H. Chen, D. Zhao, Q. Chen, and R. Chai, "Joint computation offloading and radio resource allocations in small-cell wireless cellular networks," *IEEE Transactions on Green Communications and Networking*, vol. 4, no. 3, pp. 745–758, 2020.

[17] J. Moysen and L. Giupponi, "A reinforcement learning based solution for self-healing in LTE networks," in *2014 IEEE 80th Vehicular Technology Conference (VTC2014-Fall)*, pp. 1–6, IEEE, 2014.

[18] M. Z. Asghar, F. Ahmed, and J. Hämäläinen, "Artificial intelligence enabled self-healing for mobile network automation," in *2021 IEEE Globecom Workshops (GC Wkshps)*, pp. 1–6, IEEE, 2021.

[19] T. Reshmi and M. Azath, "Improved self-healing technique for 5g networks using predictive analysis," *Peer-to-Peer Networking and Applications*, vol. 14, no. 1, pp. 375–391, 2021.

[20] M. S. Riaz, H. N. Qureshi, U. Masood, A. Rizwan, A. Abu-Dayya, and A. Imran, "Deep learning-based framework for multi-fault diagnosis in self-healing cellular networks," in *2022 IEEE Wireless Communications and Networking Conference (WCNC)*, pp. 746–751, IEEE, 2022.

[21] M. A. Hossain, A. R. Hossain, and N. Ansari, "AI in 6G: Energy-efficient distributed machine learning for multilayer heterogeneous networks," *IEEE Network*, vol. 36, no. 6, pp. 84–91, 2022.

[22] S. K. G. Peesapati, M. Olsson, S. Andersson, C. Qvarfordt, and A. Dahlen, "AI-assisted multi-operator ran sharing for energy-efficient networks," *Telecom*, vol. 4, pp. 334–368, MDPI, 2023.

[23] E. Gures, I. Shayea, M. Ergen, M. H. Azmi, and A. A. El-Saleh, "Machine learning-based load balancing algorithms in future heterogeneous networks: A survey," *IEEE Access*, vol. 10, pp. 37689–37717, 2022.

[24] G. Alsuhli, K. Banawan, K. Attiah, A. Elezabi, K. Seddik, A. Gaber, M. Zaki, and Y. Gadallah, "Mobility load management in cellular networks: A deep reinforcement learning approach," *IEEE Transactions on Mobile Computing*, vol. 22, pp. 1581–1598, 2021.

[25] W. K. Saad, I. Shayea, A. Alhammadi, M. M. Sheikh, and A. A. El-Saleh, "Handover and load balancing self-optimization models

in 5G mobile networks," *Engineering Science and Technology, an International Journal*, vol. 42, p. 101418, 2023.

[26] A. Alhammadi, W. H. Hassan, A. A. El-Saleh, I. Shayea, H. Mohamad, and W. K. Saad, "Intelligent coordinated self-optimizing handover scheme for 4G/5G heterogeneous networks," *ICT Express*, vol. 9, no. 2, pp. 276–281, 2023.

[27] D. Chalmers and M. Sloman, "A survey of quality of service in mobile computing environments," *IEEE Communications Surveys*, vol. 2, no. 2, pp. 2–10, 1999.

[28] J. Liu and O. W. Yang, "Using fuzzy logic control to provide intelligent traffic management service for high-speed networks," *IEEE Transactions on Network and Service Management*, vol. 10, no. 2, pp. 148–161, 2013.

[29] Y. Fu, S. Wang, C.-X. Wang, X. Hong, and S. McLaughlin, "Artificial intelligence to manage network traffic of 5G wireless networks," *IEEE Network*, vol. 32, no. 6, pp. 58–64, 2018.

[30] J. Shen, T. Zhou, K. Wang, X. Peng, and L. Pan, "Artificial intelligence inspired multi-dimensional traffic control for heterogeneous networks," *IEEE Network*, vol. 32, no. 6, pp. 84–91, 2018.

[31] A. Chen, J. Law, and M. Aibin, "A survey on traffic prediction techniques using artificial intelligence for communication networks," *Telecom*, vol. 2, pp. 518–535, MDPI, 2021.

[32] F. Morales, M. Ruiz, L. Gifre, L. M. Contreras, V. López, and L. Velasco, "Virtual network topology adaptability based on data analytics for traffic prediction," *Journal of Optical Communications and Networking*, vol. 9, no. 1, pp. A35–A45, 2017.

[33] M. Singh, J. Křížz, M. Kamruzzaman, V. Dhasarathan, A. Sharma, and S. A. Abd El-Mottaleb, "Design of a high-speed OFDM-SAC-OCDMA-based FSO system using EDW codes for supporting 5G data services and smart city applications," *Frontiers in Physics*, vol. 10, p. 934848, 2022.

[34] A. Sharma and J. Malhotra, "Evaluating the effects of material reflectivity and atmospheric attenuation on photonic radar performance in free space optical channels," *Journal of Optical Communications*, 2023.

[35] A. Sharma, V. Kumar, and V. Gupta, "A review on inter-satellite optical wireless communication," *International Journal of Computer Applications*, vol. 180, pp. 13–17, 2018.

[36] H. C. Leung, C. S. Leung, E. W. Wong, and S. Li, "Extreme learning machine for estimating blocking probability of bufferless OBS/OPS networks," *Journal of Optical Communications and Networking*, vol. 9, no. 8, pp. 682–692, 2017.

[37] C. Vinchoff, N. Chung, T. Gordon, L. Lyford, and M. Aibin, "Traffic prediction in optical networks using graph convolutional generative adversarial networks," in *2020 22nd International Conference on Transparent Optical Networks (ICTON)*, pp. 1–4, IEEE, 2020.

[38] M. F. A. Fauzi, R. Nordin, N. F. Abdullah, H. A. Alobaidy, and M. Behjati, "Machine learning-based online coverage estimator (MLOE): Advancing mobile network planning and optimization," *IEEE Access*, vol. 11, pp. 3096–3109, 2023.

[39] L. Dai, H. Zhang, and Y. Zhuang, "Propagation-model-free coverage evaluation via machine learning for future 5G networks," in *2018 IEEE 29th Annual International Symposium on Personal, Indoor and Mobile Radio Communications (PIMRC)*, pp. 1–5, IEEE, 2018.

[40] S. Mohammadjafari, S. Roginsky, E. Kavurmacioglu, M. Cevik, J. Ethier, and A. B. Bener, "Machine learning-based radio coverage prediction in urban environments," *IEEE Transactions on Network and Service Management*, vol. 17, no. 4, pp. 2117–2130, 2020.

[41] H. Mahboubi, S. Blouin, and A. G. Aghdam, "A machine learning assisted method for coverage optimization in a network of mobile sensors," *IEEE Transactions on Industrial Informatics*, vol. 19, pp. 7301–7311, 2022.

[42] F. Lyu, C. Cheng, J. Zhu, X. Cheng, L. Xu, Z. Wang, J. Qiao, L. Liu, and Z. Di, "Coverage estimation of mobile network using supervised learning model on artificial estimation dataset," in *2021 International Conference on Information and Communication Technologies for Disaster Management (ICT-DM)*, pp. 214–220, IEEE, 2021.

[43] P. V. Klaine, M. A. Imran, O. Onireti, and R. D. Souza, "A survey of machine learning techniques applied to self-organizing cellular networks," *IEEE Communications Surveys & Tutorials*, vol. 19, no. 4, pp. 2392–2431, 2017.

[44] A. Mahmood, M. L. M. Kiah, M. R. Z'aba, A. N. Qureshi, M. S. S. Kassim, Z. H. A. Hasan, J. Kakarla, I. S. Amiri, and S. R. Azzuhri, "Capacity and frequency optimization of wireless backhaul

network using traffic forecasting," *IEEE Access*, vol. 8, pp. 23264–23276, 2020.

[45] A. Abdelmoaty, D. Naboulsi, G. Dahman, G. Poitau, and F. Gagnon, "Resilient topology design for wireless backhaul: A deep reinforcement learning approach," *IEEE Wireless Communications Letters*, vol. 11, no. 12, pp. 2532–2536, 2022.

[46] Frost & Sullivan, "Global growth opportunities for telecommunications service providers in artificial intelligence (AI)," 2023. www.marketresearch.com/Frost-Sullivan-v383/Global-Growth-Opportunities-Telecommunications-Service-34113479

[47] Ericsson, "Improving mobile communications infrastructure through AI: AI4Mobile project boosts performance for industry and transport," 2023. www.ericsson.com/en/news/3/2023/improving-mobile-communications-infrastructure-through-ai-ai4mobile-project-boosts-performance-for-industry-and-transport

[48] X. Sun, C. Zhou, X. Duan, and T. Sun, "A digital twin network solution for end-to-end network service level agreement (SLA) assurance," *Digital Twin*, vol. 1, p. 5, 2021.

[49] J. Deng, Q. Zheng, G. Liu, J. Bai, K. Tian, C. Sun, Y. Yan, and Y. Liu, "A digital twin approach for self-optimization of mobile networks," in *2021 IEEE Wireless Communications and Networking Conference Workshops (WCNCW)*, pp. 1–6, IEEE, 2021.

[50] Y. Wang, Z. Su, S. Guo, M. Dai, T. H. Luan, and Y. Liu, "A survey on digital twins: Architecture, enabling technologies, security and privacy, and future prospects," *IEEE Internet of Things Journal*, vol. 10, no. 17, pp. 14965–14987, 2023.

[51] H. Ahmadi, A. Nag, Z. Khar, K. Sayrafian, and S. Rahardja, "Networked twins and twins of networks: An overview on the relationship between digital twins and 6G," *IEEE Communications Standards Magazine*, vol. 5, no. 4, pp. 154–160, 2021.

[52] E. Li, L. Zeng, Z. Zhou, and X. Chen, "Edge AI: On-demand accelerating deep neural network inference via edge computing," *IEEE Transactions on Wireless Communications*, vol. 19, no. 1, pp. 447–457, 2019.

[53] T. X. Tran, A. Hajisami, P. Pandey, and D. Pompili, "Collaborative mobile edge computing in 5G networks: New paradigms, scenarios,

and challenges," *IEEE Communications Magazine*, vol. 55, no. 4, pp. 54–61, 2017.

[54] H. Fang, X. Wang, and S. Tomasin, "Machine learning for intelligent authentication in 5G and beyond wireless networks," *IEEE Wireless Communications*, vol. 26, no. 5, pp. 55–61, 2019.

[55] J. Ali-Tolppa, S. Kocsis, B. Schultz, L. Bodrog, and M. Kajo, "Self-healing and resilience in future 5G cognitive autonomous networks," in *2018 ITU Kaleidoscope: Machine Learning for a 5G Future (ITU K)*, pp. 1–8, IEEE, 2018.

[56] A. Krayani, M. Farrukh, M. Baydoun, L. Marcenaro, Y. Gao, and C. S. Regazzoni, "Jammer detection in M-QAM-OFDM by learning a dynamic Bayesian model for the cognitive radio," in *2019 27th European Signal Processing Conference (EUSIPCO)*, pp. 1–5, IEEE, 2019.

[57] J. Qiu, L. Du, Y. Chen, Z. Tian, X. Du, and M. Guizani, "Artificial intelligence security in 5G networks: Adversarial examples for estimating a travel time task," *IEEE Vehicular Technology Magazine*, vol. 15, no. 3, pp. 95–100, 2020.

CHAPTER 4

Application of Evolutionary Algorithms on Wireless Sensor Network Coverage Optimization

Comparative Study

Amir Parnianifard, Abhishek Sharma, and Sushank Chaudhary

4.1 INTRODUCTION

Network coverage, which is essential for collecting location information and delivering perception services, is one of the main issues with wireless sensor networks (WSNs) [1]. Sensor networks typically contain a huge number of nodes and operate in complex environments that humans cannot access. Nodes are frequently dispersed at random by aircraft pitching. However, doing so may result in too many or too few nodes in some places, creating blind or perception-overlapping zones. The location of the nodes must thus be adjusted dynamically using an optimization technique [2]. In WSN, coverage is frequently used as a metric for service quality. From this vantage point, coverage refers to how effectively the sensors monitor their intended area [3]. The literature has given a lot of attention to the

DOI: 10.1201/9781003472506-4

WSN coverage optimization problem by maximizing the coverage of the monitoring area [1,4–6].

Ensuring sufficient coverage while adhering to specific application demands frequently presents NP-hard optimization challenges. These complexities arise due to the need to strategically deploy sensor nodes in a manner that maximizes coverage efficiency while considering constraints such as energy consumption, network connectivity, and environmental factors [7–9]. As the number of sensors to be deployed increases, the search space expands exponentially, i.e., if assume m represents the number of allowable locations for each sensor, then the search space for n sensors can be m^n possible points [10]. Many researchers have explored the application of metaheuristic techniques, also known as evolutionary algorithms, to address the coverage optimization problem, achieving significant advancements [11].

These methodologies encompass a variety of optimization techniques used in computational intelligence. Genetic algorithm (GA) [12] employs principles of natural selection to iteratively evolve solutions through operations like selection, crossover, and mutation, simulating biological evolution. Particle swarm optimization (PSO) [13] emulates the collective behavior of birds or fish to iteratively refine solutions based on individual and group experiences within the swarm. Ant colony optimization (ACO) [14] draws inspiration from ants' foraging behavior, utilizing pheromone trails to find optimal paths and solve complex problems efficiently. Ant lion optimization (ALO) [15] replicates the predatory strategy of ant lions, leveraging trapping mechanisms to explore and exploit search spaces effectively. Differential evolution (DE) [16] is a population-based method that enhances solution quality and convergence speed by iteratively evolving solutions using differences between randomly selected pairs. Finally, the grey wolf optimizer (GWO) [17], a newer swarm intelligence algorithm, mimics grey wolves' hunting tactics and hierarchical social structure to balance exploration and exploitation in search of optimal solutions.

Generally, these algorithms are widely applicable to a range of critical optimization challenges in WSNs. For instance, they are used in localization problems [18–21], where the goal is to accurately determine the positions of sensor nodes within the network. They are also employed in designing minimum energy topologies [22], which aim to minimize energy consumption while maintaining network connectivity and performance. Another vital application pertains to optimizing the shortest path for data transmission [23,24]. This optimization objective aims to minimize both the travel costs associated with the sink and the time taken for data collection, thereby significantly enhancing the efficiency and effectiveness of data-gathering processes in WSNs. This optimization task is pivotal for ensuring timely and cost-effective data retrieval across vast sensor network deployments. Additionally, these algorithms are utilized to maximize signal coverage [25,26], ensuring that the monitoring area is adequately covered by the sensor nodes to detect events and transmit information reliably.

However, this chapter specifically hones in on the common WSN coverage optimization problems [1,4], addressing the challenge of determining the optimal placement of sensor nodes to maximize the coverage area while minimizing the number of sensors used. This focus is crucial because effective coverage directly impacts the network's ability to monitor and report on environmental conditions accurately and efficiently. The study explores various metaheuristic algorithms to solve these coverage problems, providing a comprehensive comparison and analysis of their performance in different scenarios.

This study aims to conduct a structured comparative analysis to assess the performance of the four most prevalent metaheuristic techniques—PSO, GWO, GA, and DE—in addressing practical WSN coverage optimization challenges. The study conducts a statistical evaluation of the results generated by these optimizers, focusing on three main criteria:

- **Accuracy:** This criterion measures how close the optimizer's solutions are to the best-known or true optimal solution. High accuracy indicates that the optimizer consistently finds solutions that are near the optimal value, demonstrating its effectiveness in solving the given problem.
- **Robustness:** Robustness assesses the optimizer's ability to produce consistent results across multiple runs. An optimizer is considered robust if it delivers reliable and stable performance despite variations in initial conditions or stochastic elements in the optimization process. This is typically evaluated using the standard deviation of the fitness values obtained from multiple iterations.
- **Computational Time:** This metric evaluates the total time required by the optimizer to converge to the optimal solution. It includes the time taken for all iterations and function evaluations. An efficient optimizer should minimize computational time while still providing

accurate and robust solutions, making it suitable for real-time or large-scale applications.

Our contributions are twofold: (i) we systematically compare and evaluate these four well-established metaheuristic techniques commonly used in WSN coverage optimization, and (ii) we offer practical guidance to assist practitioners in selecting the most suitable method based on considerations of accuracy, robustness, and efficiency in addressing large-scale coverage optimization problems involving 30–60 design variables associated with sensor set coordinates.

This chapter is structured as follows. Section 4.2 reviews the literature, including recent studies about metaheuristic-based coverage optimization in WSN. The basic methodology of each participated optimizer in the current comparative study is explained in Section 4.3. The statement of the problem behind coverage WSN optimization is mentioned in Section 4.4. Section 4.5 includes the specification of benchmark functions and simulation setup required for systematic comparison of algorithms and implementation of algorithms in benchmark cases. The obtained comparison results are also statistically analyzed and discussed in this section. Finally, this chapter is concluded in Section 4.6.

4.2 RELATED WORKS

To enhance design quality and reduce the design cycle duration, optimization technologies are increasingly being applied to engineering design challenges. Optimization typically involves identifying the optimal levels of a set of design variables, X, to achieve one or more objectives, $f(X)$, while adhering to constraints on the design variables, $g(X)$. These constraints, which can be expressed as equalities or inequalities, restrict the design space and complicate the search for the optimal solution. Despite these challenges, the general framework of a mathematical programming model can be described as follows:

$$\begin{aligned} &\text{Min or Max}: f(X) \\ &\text{Subject to}: g_j(X) \leq 0, \quad j = 1, 2, \ldots, J \end{aligned} \tag{4.1}$$

The objective function $f(X)$ and the inequality constraints $g_j(X)$, where X represents the decision variables, form the basis of most engineering design problems [27]. Since engineering designs typically involve constraints [28–30], effective methods are essential to handle these constraints robustly. Optimization approaches that aim to find optimal solutions while satisfying these constraints are broadly categorized into model-based and metamodel-based (or surrogate-based) strategies [31–35] and [36].

Classical model-based techniques include metaheuristic methods such as GA, PSO, DE, and GWO [37]. These methods are favored not only by computer scientists but also by researchers across diverse disciplines due to their straightforward implementation, adaptability, lack of derivative requirements, and capability to escape local optima [17].

In scenarios involving "expensive optimization" problems, like those encountered in computer-aided design or simulation-based optimization, researchers often prefer metamodel-based approaches that integrate surrogates with metaheuristics [38–43]. These surrogate-based methods allow for efficient exploration of the design space by replacing computationally expensive objective functions with surrogate models, thereby reducing the overall computational burden. The scope of the current chapter is confined to the application of model-based metaheuristic methods, which have shown significant advantages in addressing various WSN optimization problems [20,44,45].

The coverage rate is a fundamental metric in numerous applications of WSNs as it determines the effectiveness and extent of the monitoring area. The deployment strategy of sensor nodes plays a critical role in this regard, as efficient node placement can significantly enhance both energy efficiency and the overall quality of network services [46,47]. In many scenarios, sensor nodes are distributed randomly within a given workspace, which often leads to an excessive number of nodes. This overabundance can reduce both the coverage and the quality of monitoring due to overlaps and redundancy in the sensor network [47,48]. Therefore, optimizing the placement of sensors is essential to maximize the coverage ratio of WSNs in practical operational environments.

The challenge of finding the optimal node placement strategy becomes particularly complex when dealing with large-scale sensor networks, where the number of nodes and the area to be covered are substantial [49]. In this context, traditional methods may fall short due to their inability to efficiently handle the combinatorial nature of the problem. As a result, researchers have increasingly turned to metaheuristic algorithms to address these optimization challenges, leading to significant advancements in WSN coverage.

Metaheuristic techniques, known for their ability to find near-optimal solutions to complex problems, have been extensively studied and applied in WSNs. For instance, a novel adaptation of the fruit fly optimization algorithm was proposed by Song et al. [50], which introduced a new change step specifically aimed at enhancing coverage. This variant showed improved performance in optimizing the placement of sensor nodes compared to traditional methods. Similarly, the ACO algorithm has been applied by Liao et al. [51] to expand coverage areas. The ACO algorithm, inspired by the foraging behavior of ants, proved effective in finding optimal paths and positions for sensor nodes, thereby maximizing the coverage.

Another notable contribution is from Ari et al., who utilized the bacterial foraging optimization technique [52] to improve coverage rates. This method, inspired by the foraging patterns of bacteria, was successful in optimizing sensor placement by simulating bacterial movement and interaction, which led to better coverage performance. Each of these metaheuristic approaches demonstrates the potential and versatility of such algorithms in solving the complex problem of sensor deployment in WSNs.

Furthermore, the application of these algorithms is not limited to theoretical improvements. Practical implementations have shown that metaheuristic methods can be effectively used in real-world WSN deployments to enhance monitoring capabilities. The continuous development and refinement of these algorithms ensure that WSNs can achieve higher coverage rates, better energy efficiency, and improved service quality, which are critical for various applications such as environmental monitoring, security surveillance, and industrial automation, see [1,4–6].

Vilovic et al. [53] utilized the ACO method to optimize transmitter positions with the goal of maximizing the average received power. This approach involves simulating the behavior of ants to find optimal paths, which in this context translates to finding the best positions for transmitters to ensure maximum signal strength. Comparative studies have shown that while PSO and GA yield similar antenna positions, ACO tends to require a greater amount to fitness function evaluations. This results in slower convergence times for ACO compared to PSO and GA, as detailed in [10,54,55]. Despite this, GA has been notably effective in the design of high-performance antennas, a success attributed to its robust search capabilities and adaptability, which are common traits among most metaheuristic algorithms.

Additionally, PSO has been employed to determine optimal positions for base stations to enhance signal coverage, as noted in Yangyang et al. [56]. This method leverages the social behavior of particles, or agents, to iteratively improve the positions of base stations, thereby ensuring optimal coverage with efficient computation. The versatility and effectiveness of PSO make it a popular choice for various optimization problems in wireless communications.

The GWO, introduced by Mirjalili et al. [17], has rapidly gained traction in the field of WSNs. GWO simulates the leadership hierarchy and hunting behavior of grey wolves in nature to search for optimal solutions. This algorithm has demonstrated strong capabilities in achieving optimal coverage solutions for WSNs, as supported by the findings by Miao et al. [57]. The hierarchical approach of GWO, which involves alpha, beta, and delta wolves guiding the search, has proven effective in balancing exploration and exploitation during the optimization process.

These advancements highlight the significant contributions of metaheuristic algorithms in the field of wireless communications. Each algorithm, with its unique mechanisms and strategies, offers distinct advantages. The choice of algorithm often depends on the specific requirements of the problem at hand, such as the need for faster convergence or higher solution accuracy. As research in this area continues to evolve, these algorithms are expected to become even more refined and capable, further enhancing the performance and reliability of WSNs and other wireless communication systems.

4.3 BEYOND METAHEURISTICS PARTICIPATED IN THE CURRENT WORK

In this section, the basic concepts and terminology of each of the four optimizers which participated in the comparative study in this work, e.g., PSO, GWO, GA, and DE are explained.

4.3.1 Particle Swarm Optimization

Reference [13] introduced the standard PSO algorithm, which models the collective behavior observed in social animals such as fish schools and bird flocks. This algorithm utilizes a population of particles, each representing a potential solution to the optimization problem. The critical parameters in

PSO encompass the quantity of particles, the positional coordinates, and velocities of each particle within the solution space. Additionally, the neighborhood structure defines the communication patterns among particles, influencing their interactions and collective behavior during the optimization process. This topology communication plays a pivotal role in determining how particles exchange information, adapt their movements, and converge toward optimal solutions by leveraging both personal and neighborhood best positions. The PSO algorithm begins by initializing the population, with each particle assigned a random position and velocity within the solution space. Each particle's position represents a possible solution, and its velocity indicates the direction and magnitude of its movement. The algorithm then evaluates the fitness values of all particles based on a predefined fitness function that quantifies how well a given solution solves the problem.

After evaluating fitness values, PSO updates the individual best position (p_best) for each particle, which is the best solution the particle has encountered. Additionally, it updates the global best position (g_best), which is the best solution found by any particle in the swarm. These updates guide the particles toward more promising regions of the solution space.

The velocity and position of each particle are then updated using the following equations:

$$v_{id}^{t+1} = v_{id}^{t} + c_1 \cdot \text{r and } (0,1) \cdot \left(p_{id}^{t} - x_{id}^{t}\right) + c_2 \cdot \text{ r and } (0,1) \cdot \left(p_{gd}^{t} - x_{id}^{t}\right) \quad (4.2)$$

$$\text{and } x_{id}^{t+1} = x_{id}^{t} + v_{id}^{t}$$

In Eq. (4.2), the terms are defined as follows:

- v_{id}^{t} : Velocity of particle i in dimension d at iteration t .
- x_{id}^{t} : Position of particle i in dimension d at iteration t .
- p_i : Best position found so far by particle i .
- p_g : Best position found so far among the particle's neighbors.
- c_1 : Scaling factor (often referred to as cognitive coefficient) governing the influence of p_i on the particle's movement.
- c_2 : Scaling factor (often referred to as social coefficient) governing the influence of p_g on the particle's movement.
- $\text{rand}(0,1)$: Random number uniformly distributed between 0 and 1.

The exploration-exploitation balance in PSO is determined by comparing the difference between p_i and x_{id}^t and between p_g and x_{id}^t:

- Exploration takes place when a considerable disparity exists between the best individual position (p_i) and the current position of particle i (x_{id}^t), or between p_g (best neighbor position) and x_{id}^t. This phase involves particles exploring new areas of the solution space to discover potentially better solutions beyond their current positions. It enables the algorithm to search for global optima by encouraging particles to venture into less explored regions, thereby enhancing the diversity of the search and preventing premature convergence to local optima.
- Exploitation occurs when there is a small difference between these values, indicating that the particle is exploiting the local or global best positions.

These principles guide the movement of particles in the search space toward better solutions over successive iterations in the PSO algorithm. The PSO algorithm effectively combines local and global search strategies by leveraging the social behavior of particles. Its simplicity, ease of implementation, and ability to find high-quality solutions make it a popular choice for solving a wide range of optimization problems.

4.3.2 Grey Wolf Optimizer

The original GWO algorithm, introduced by Mirjalili et al. [17], represents a significant advancement in the field of swarm intelligence-based algorithms. The GWO derives its principles from the intricate social structures and hunting dynamics exhibited by grey wolves in their natural environment. In this framework, the population of potential solutions is organized according to a strict hierarchy. The most optimal solution within the population is designated as the alpha (α) wolf, which leads the hunt and makes the most crucial decisions. The second and third most optimal solutions are termed beta (β) and delta (δ) wolves, respectively. These wolves assist the alpha in decision-making and strategy formation. All other solutions in the population are categorized as omega (ω) wolves, which follow the leaders and complete the hierarchy. This hierarchical structure allows the GWO to balance exploration and exploitation effectively during the optimization process. The algorithm mimics the grey wolves' natural hunting process, including tracking, chasing, and attacking prey, to converge toward the optimal solution. This unique approach enables the GWO to perform robustly across various optimization problems.

In the hunting process, grey wolves simulate the encircling of their prey, which is a key aspect of their strategy for capturing food. This behavior is mathematically modeled in the GWO algorithm by the following equation:

$$\vec{X}(t+1)=\vec{X}_p(t)-\vec{A}\cdot\left|\vec{C}\cdot\vec{X}_p(t)-\vec{X}(t)\right| \tag{4.3}$$

In this context, t represents the existing iteration index, coefficient vectors indicate with $\vec{A}$ and $\vec{C}$, and the position vector of the prey with $\vec{X}_p$. The expression $\vec{X}$ denotes the position vector of a grey wolf. The encircling mechanism enables grey wolves to gradually approach and converge on their target (optimal solution) during the optimization process.

The coefficient vectors $\vec{A}$ and $\vec{C}$ are calculated using the following equations:

$$\vec{A}=2\vec{a}\cdot\vec{r}_1-\vec{a} \text{ and } \vec{C}=2\vec{r}_2 \tag{4.4}$$

Here, r_1 and r_2 are random vectors with components drawn from the interval $\left[0,1\right]$. The vector $\vec{a}$ a linearly decreases from 0 over the course of the iterations, following the equation:

$$\vec{a}(t)=2-\frac{2t}{\text{MaxIter}} \tag{4.5}$$

where MaxIter denotes the total number of iterations. This decrement helps balance exploration and exploitation by reducing the influence of the randomness as the algorithm progresses, thereby allowing the wolves to converge toward the optimal solution.

The grey wolves update their positions based on the positions of the alpha (α), beta (β), and delta (δ) wolves, which are the best solutions found so far. This hierarchical update strategy is critical to the GWO's effectiveness, ensuring that the search process is guided by the best solutions at each iteration. The updated positions of the wolves are influenced by the positions of these top three wolves, which act as guides in the search space.

This mechanism is designed to simulate the natural social hierarchy of grey wolves besides the collaborative hunting strategy, enabling the algorithm to efficiently explore and exploit the search space to find optimal solutions.

The grey wolves update their positions based on the positions of α, β, and δ as follows:

$$\vec{X}(t+1)=\frac{1}{3}\begin{pmatrix} \vec{X}_{\alpha}(t)-\vec{A}_{1}\cdot\left|\vec{C}_{1}\cdot\vec{X}_{\alpha}(t)-\vec{X}(t)\right| \\ +\vec{X}_{\beta}(t)-\vec{A}_{2}\cdot\left|\vec{C}_{2}\cdot\vec{X}_{\beta}(t)-\vec{X}(t)\right| \\ +\vec{X}_{\delta}(t)-\vec{A}_{3}\cdot\left|\vec{C}_{3}\cdot\vec{X}_{\delta}(t)-\vec{X}(t)\right| \end{pmatrix} \tag{4.6}$$

where $\vec{A}$ and $\vec{C}$ are obtained as in Eq. (4.4), see [17, 57].

4.3.3 Genetic Algorithm

A genetic algorithm, inspired by Darwin's theory of natural selection [58,59], starts by selecting the fittest individuals from a population. These individuals reproduce, passing on their traits to the next generation. Offspring of fitter parents perform better, repeating the process to eventually yield the fittest generation. This approach applies to solving search problems by evaluating potential solutions and selecting the best ones. The genetic algorithm involves:

- **Initial Population:** Individuals represent potential solutions encoded as binary strings of genes.
- **Fitness Function**: Evaluates an individual's competitiveness based on assigned fitness scores.
- **Selection:** Fittest individuals are chosen for reproduction.
- **Crossover:** Genes are exchanged between selected parents to create offspring.
- **Mutation:** Occasional random changes maintain genetic diversity and prevent premature convergence.

4.3.4 Differential Evolution

DE was first conceptualized by Storn [60], who also published the first publication on the topic as a technical report. By then, DE had become a popular choice among academics for addressing optimization challenges across various fields [61]. Startup and evolution are the two phases of DE's functioning. First, a population is formed at random. Then, during the evolution phase, the generated population is repeatedly subjected to the processes of mutation, crossover, and selection until a termination condition is met [62]. The algorithmic procedure of DE can be summarized as follows. Individuals are selected randomly from intervals $\left[x_j^L, x_j^U\right]$, where x_j^L and x_j^U represent the lower and upper bounds of decision parameters. Each of these selected individuals then undergoes mutation, recombination, and subsequent selection. Mutation plays a crucial role in expanding the search space. For any given individual u, three distinct parameters i, r_1, and r_2 and r_3 are randomly chosen. The donor vector r_{i+1} is computed as $r_{i+1} = r_1 + \alpha(r_2 - r_3)$, where α, termed the mutation factor, ranges between 0 and 2. This process enhances diversity and exploration within the population. These three parameters are replaced to selected individual u if the $\mathrm{r\ and}() > \mathrm{CR}$, where CR is a constant probability in $U[0,1]$. If the fitness function of the updated individual is less than the original individual u, then u is removed from the set of individuals, and an updated one is replaced. Until a stopping condition is met, mutation, recombination, and selection proceed.

4.4 STATEMENT OF WSN COVERAGE OPTIMIZATION

The sensor node's coverage area is defined as a circular region with a constant radius, determining the inclusion of locations within its operational scope. This metric is essential for evaluating how effectively a WSN monitors a two-dimensional (2D) area. To quantify the network's coverage rate across this monitoring zone, the entire space is systematically divided into $M \times N$ pixels. The coverage rate itself is defined as the ratio of covered points (x) to the total number of pixels ($M \times N$), providing a measure of how comprehensively the WSN can detect and monitor activity within the designated area. This evaluation assumes uniform sensor characteristics throughout and simplifies the analysis by disregarding potential influences from the boundaries of the monitored region.

In a monitoring area with N sensor nodes, each with a perceived radius R, denoted as $T = \{t_1, t_2, \ldots, t_N\}$, the coordinates of the i-th sensor node

t_i are (x_i, y_i). Similarly, the coordinates of the j-th monitoring node h are x_j, y_j. The distance between h and t_i is expressed by the following equation:

$$d(t_i - h) = \left((x_i - x_j)^2 + (y_i - y_j)^2 \right)^{\frac{1}{2}} \tag{4.7}$$

This study adopts a binary perception paradigm, where the likelihood of sensor t_i detecting monitoring node h is quantified by the following probability expression:

$$p(t_i - h) = \begin{cases} 1, & d(t_i - h) \le 0 \\ 0, & d(t_i - h) > 0 \end{cases} \tag{4.8}$$

Sensor t_i effectively covers monitoring point h when $p(t_i - h) = 1$. Conversely, if $p(t_i - h) = 0$, sensor t_i does not cover h. Given that multiple sensor nodes may concurrently observe the same position within the monitoring region, the probability that monitoring point h is detected by the node set T can be expressed as:

$$p(T,h) = 1 - \prod_{t_i \in T} \left(1 - \left(p(t_i - h)\right)\right) \tag{4.9}$$

The percent coverage rate of the node set T over the entire monitoring area is:

$$R_c = \frac{\sum_{h \in M \times N} p(T,h)}{M \times N} \times 100 \tag{4.10}$$

In this chapter, to test optimizers in a more challenging design space, we assume some obstacle areas over the whole design area $M \times N$. Figure 4.1 shows the schematic 2D map (100×100) of the test area used in the benchmark functions employed in the current comparative study.

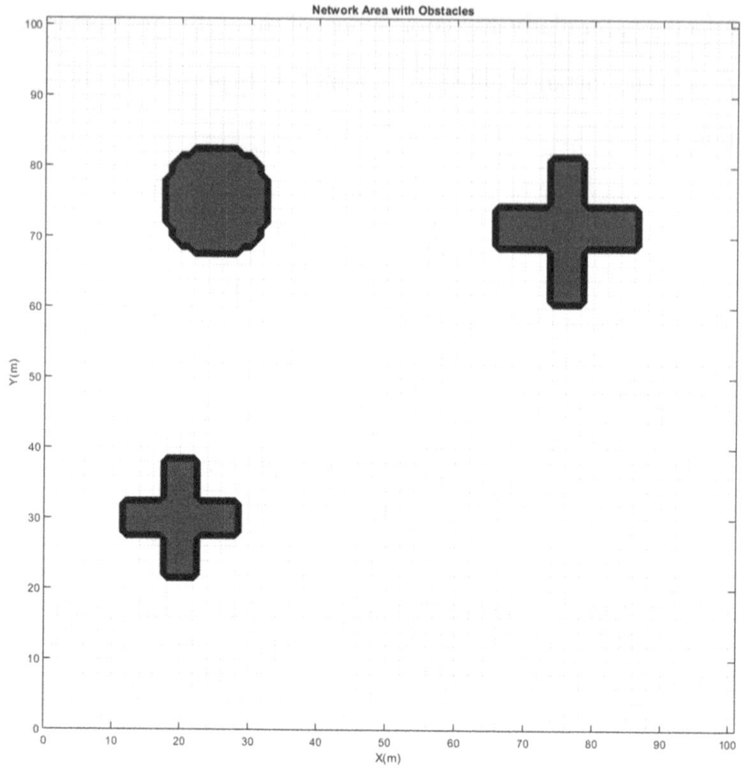

FIGURE 4.1 Schematic representation of network area with obstacles. The free pixels (not including the obstacles' coordinates) are considered a network area that needs to be covered by sensor nodes and design space in optimization models.

Let's assume the obstacle areas are not interested to be covered by sensor nodes, and just $M \times N - obstacles$ are targeted to be covered. Therefore, Eq. (4.10) can be replaced by:

$$R_c = \frac{\sum_{h \in Mf \times Nf} p(T,h)}{Mf \times Nf} \times 100 \tag{4.11}$$

where $Mf \times Nf$ are coordinates (pixels) of free area in XY direction (2D) map that does not include pixels of obstacles' areas. Here, the fitness function for the WSN coverage optimization problem is considered based on the formulation in Eq. (4.11).

4.5 SYSTEMATIC COMPARISON OF OPTIMIZERS

4.5.1 Simulation Setup

In this study, six distinct cases of WSN coverage optimization problems are investigated to compare different optimizers, as detailed in Table 4.1.

These cases vary in terms of the number of sensor nodes and the maximum number of function evaluations tailored for each optimizer. In the 2D monitoring area, the search agent's dimension (number of design variables) doubles the number of sensor nodes along both the *X*-direction and *Y*-direction. For instance, cases with 15 sensor nodes involve 30 design variables in the first, second, and third cases, and 60 variables in the fourth, fifth, and sixth cases.

To ensure optimal convergence within the specified number of function evaluations, the maximum number of iterations for each optimizer is adjusted accordingly. For example, the maximum iterations initially set for the optimizers based on the number of function evaluations are 5, 10, 20, 5, 10, and 20 for cases 1–6 in Table 4.1, respectively.

During the initialization phase, the initial coordinates of all sensor nodes are set as the initial input values for the optimizers, corresponding to the initial population size. These sensor nodes are randomly distributed across the designated area, with each search agent in every algorithm representing a sensor node placement technique.

The results yielded by every optimization model comprise two crucial aspects: the optimal percentage coverage rate, which measures the extent of the monitored area effectively covered by the WSN, and the precise spatial coordinates of each sensor node deployed within the network. These outcomes are essential for evaluating the performance and effectiveness of each optimization strategy in enhancing WSN coverage capabilities. The simulations are conducted using Matlab®/R2022b to construct and simulate

TABLE 4.1 Six Different Problems Are Employed in the Comparison of Optimizers

Case No.	Number of Sensors	Max, Function Evaluations of Optimizers
1	15	150
2	15	300
3	15	600
4	30	300
5	30	600
6	30	1,200

each problem's optimization model. To ensure robust comparison results, each problem undergoes evaluation with 10 individual repetitions across the six different cases.

In these experiments, the monitoring area measures 100×100 m, with a sensor's perceived coverage radius set at 10 m, and the network area divided into 10,000 pixels. However, to add complexity and challenge the optimizers further, some pixels are designated as obstacles within the network area, as illustrated in Figure 4.1. Specifically, 1,636 pixels are obstructed, leaving 8,362 pixels monitored by the coordinates of monitoring and sensor nodes.

The fitness function for optimizers is based on the percent coverage rate of the WSN, formulated as Eq. (4.11), with the objective of maximizing this fitness function. The study focuses on evaluating the performance of four optimizers—PSO, GWO, GA, and DE—in solving each of the six optimization problems.

4.5.2 Results and Discussion

For each specific case outlined in Table 4.1 and detailed in the previous section, all four optimizers—PSO, GWO, GA, and DE—are systematically evaluated. Each optimizer is tasked with optimizing the placement of sensor nodes within the network area, aiming to maximize the percent coverage rate as the fitness function. The results obtained include the optimal positions of all sensor nodes and the corresponding optimal fitness function values.

To ensure the reliability of the findings, each optimization problem is repeated 10 times with identical conditions, allowing for a comprehensive evaluation of each optimizer's performance. The evaluation criteria encompass three key metrics: average fitness function values across all repetitions (accuracy), standard deviation (Std) of fitness function values across repetitions (robustness), and total computational time required for each optimizer across all repetitions.

This rigorous approach enables a thorough analysis of how each metaheuristic performs under various conditions and problem complexities. The results obtained from each case are meticulously examined and discussed in the subsequent sections to provide insights into the effectiveness and efficiency of each optimizer in addressing WSN coverage optimization challenges.

- **Case 1:**

In the first case, which involves optimizing the placement of 15 sensor nodes, each optimizer operates with 30 design variables corresponding to the coordinates of these sensors. This scenario is characterized by the lowest number of function evaluations allotted to each optimizer, specifically 150 evaluations. Consequently, optimizations in this case are configured with a fixed number of five iterations, and an initial population size set to 30.

Additionally, specific parameters are carefully selected for PSO. The velocity clamping factor, cognitive constant, and social constant are uniformly set to 2. The inertia weight varies dynamically between 0.4 and 0.9, ensuring a balance between exploration and exploitation throughout the optimization process.

The optimization process using each optimizer of PSO, GWO, GA, and DE has been separately conducted over ten individual repetitions. According to the obtained result, the relevant box plot, total elapsed time, and representation of obtained fitness function (coverage rate), by different optimizers over ten individual repetitions are provided in Figure 4.2.

The statistical results by different optimizers in the first case of the WSN coverage optimization problem are summarized in Table 4.2.

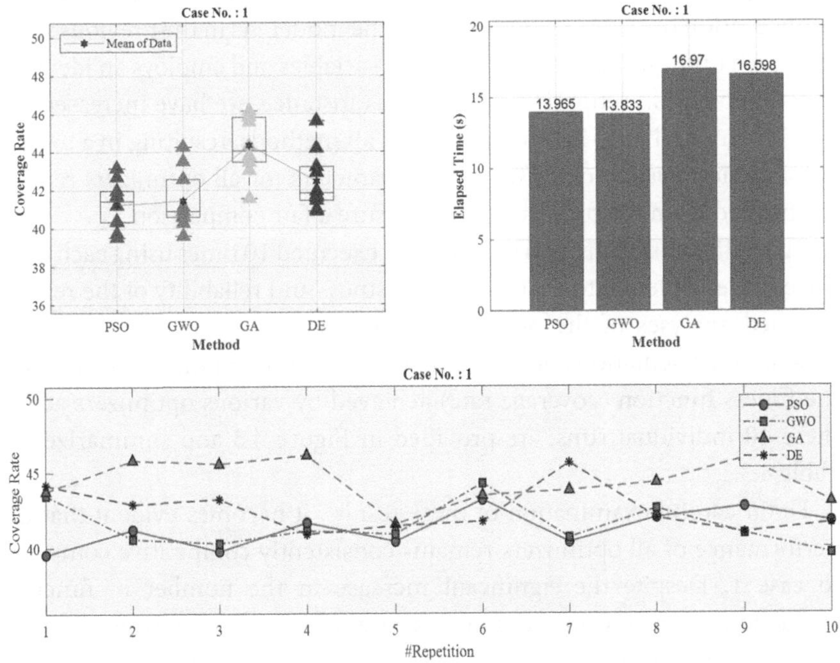

FIGURE 4.2 Box plot, total elapsed time, and representation of obtained fitness function (coverage rate) by different optimizers over 10 individual repetitions for the first case in the optimization problem of WSN coverage.

TABLE 4.2 Statistical Outcomes Achieved by Various Optimizers in the Case 1 WSN Coverage Optimization Problem

Method	Ave	Std	Max	Min	Time Elapsed (s)
PSO	41.24	1.22	43.16	39.54	13.97
GWO	41.44	1.50	44.26	39.60	13.83
GA	44.34	1.50	46.15	41.56	16.97
DE	42.45	1.53	45.62	40.89	16.60

It is apparent from this table that the GA algorithm provides a higher coverage rate followed by DE in the second rank. Both GWO and PSO provided competitive results as their fourth rank to obtain the best fitness function. Despite higher accuracy by GA, however, this algorithm elapsed much more computational time compared with other alternatives to investigate the optimal coordinates of sensor nodes. The GWO method is conducted a bit faster than the three other methods.

- **Case 2:**

In this particular case, we maintain consistency by considering the same configuration of 15 sensor nodes within the model. As in the previous scenario, each optimizer addresses 30 design variables and employs an identical initial population setup. However, in this instance, we have increased the maximum number of iterations to 10 for all methods, resulting in a total of 300 function evaluations. The other parameters for all optimizers remain unchanged from the previous case to ensure a fair comparison.

The optimization process is rigorously executed 10 times using each optimizer independently to ensure the robustness and reliability of the results. Detailed analyses of the outcomes, including comprehensive box plots, detailed representations of total elapsed time, and graphical depictions of the fitness function (coverage rate) achieved by various optimizers across these 10 individual runs, are provided in Figure 4.3 and summarized in Table 4.3.

Upon careful examination of these results, it becomes evident that the performance of all optimizers remains consistently competitive compared to case 1. Despite the significant increase in the number of function evaluations from 150 in the previous case to 300 in the current case, the observed improvement in coverage rate across all algorithms is marginal, hovering around 1%. These findings underscore the nuanced balance

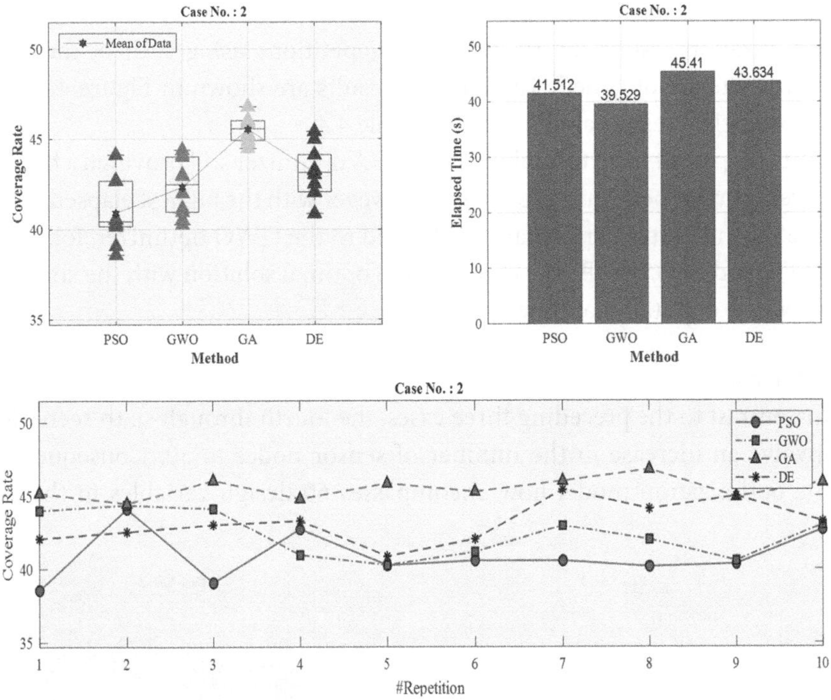

FIGURE 4.3 Box plot, total elapsed time, and representation of obtained fitness function (coverage rate) by different optimizers over 10 individual repetitions for the second case in the optimization problem of WSN coverage.

TABLE 4.3 Statistical Outcomes Achieved by Various Optimizers in the Case 2 WSN Coverage Optimization Problem

Method	Ave	Std	Max.	Min.	Time Elapsed (s)
PSO	40.90	1.74	44.12	38.57	41.51
GWO	42.34	1.56	44.38	40.27	39.53
GA	45.52	0.71	46.80	44.48	45.41
DE	43.16	1.39	45.41	40.86	43.63

between computational effort and achievable optimization gains in WSN coverage optimization problems.

- **Case 3:**

The same number of sensor nodes (15 sensors) is considered in this case; however, the allowable number of function evaluations for all optimizers amounted to 600. The same as in both previous cases, the optimization procedure is adjusted, while the maximum iteration is set to 20. The optimization results for 10 individual repetitions using each of the four optimizers are obtained. The relevant results are shown in Figure 4.4 and the statistical results are depicted in Table 4.4.

It is apparent from this table that the GA optimizer still provided a higher coverage rate (best fitness function), however with the highest elapsed time. The second better performance belonged to the GWO optimizer, followed by DE and PSO. The PSO can obtain the optimal solution with the smallest elapsed computational time.

- **Case 4:**

In contrast to the preceding three cases, the fourth through sixth scenarios involve an increase in the number of sensor nodes to 30. Consequently, the optimization model now encompasses 60 design variables in the 2D

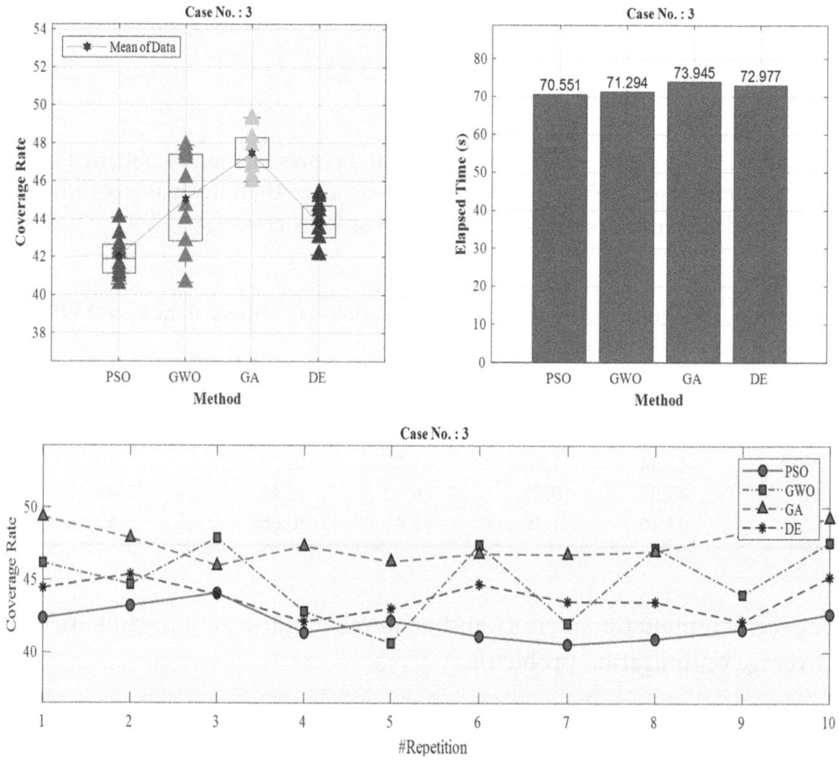

FIGURE 4.4 Box plot, total elapsed time, and representation of obtained fitness function (coverage rate) by different optimizers over 10 individual repetitions for the third case in the optimization problem of WSN coverage.

TABLE 4.4 Statistical Outcomes Achieved by Various Optimizers in the Case 3 WSN Coverage Optimization Problem

Method	Ave	Std	Max	Min	Time Elapsed (s)
PSO	42.01	1.10	44.08	40.56	70.55
GWO	45.04	2.59	47.88	40.65	71.29
GA	47.47	1.18	49.33	45.96	73.95
DE	43.80	1.16	45.37	42.12	72.98

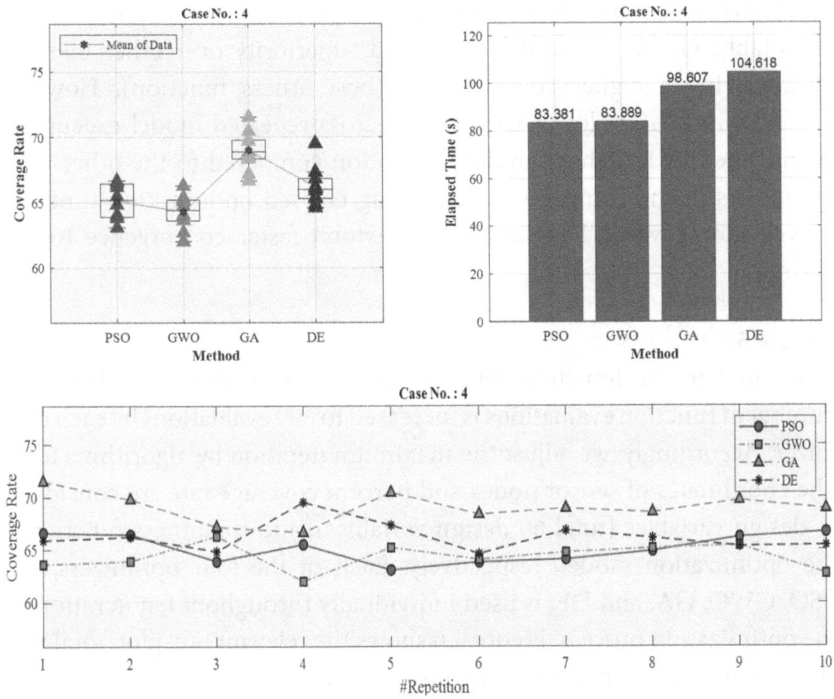

FIGURE 4.5 Box plot, total elapsed time, and representation of obtained fitness function (coverage rate) by different optimizers over 10 individual repetitions for the fourth case in the optimization problem of WSN coverage.

network area, corresponding to the coordinates of all sensors. The initial population size for each optimizer is accordingly adjusted to 60 individuals. In this case, the maximum number of iterations is set to 5 to accommodate a total of 300 function evaluations. The optimization process is conducted across 10 individual runs, and the resultant findings are graphically depicted in Figure 4.5.

TABLE 4.5 Statistical Outcomes Achieved by Various Optimizers in the Case 4 WSN Coverage Optimization Problem

Method	Ave	Std	Max	Min	Time Elapsed (s)
PSO	65.12	1.34	66.67	63.08	83.38
GWO	64.32	1.30	66.30	62.02	83.89
GA	68.98	1.46	71.51	66.65	98.61
DE	66.21	1.43	69.50	64.61	104.62

Additionally, comprehensive statistical analyses are presented in Table 4.5.

Notably, GA demonstrates continued superiority over other methods by achieving a higher coverage rate (best fitness function). However, GA exhibits slightly lower robustness across repeated model executions, as indicated by a higher standard deviation compared to the other three optimizers in the comparison. Regarding elapsed optimization time, DE proves slower, while PSO and GWO exhibit faster convergence toward optima.

- **Case 5:**

Compared to the fourth problem (previous case), in this problem, the number of function evaluations is increased to 600 evaluations by each optimizer. Accordingly, we adjust the maximum iteration by algorithms to 10. The coordinates of sensor nodes and percent coverage rate are considered as design variables (total 60 design variables), and the fitness function of the optimization model, respectively. Each of the four optimizers, e.g., PSO, GWO, GA, and DE, is used individually throughout ten iterations of the optimization process. Figure 4.6 shows the relevant box plot, total time elapsed, and representation of the fitness function (coverage rate) obtained by various optimizers over 10 distinct repetitions.

Table 4.6 also included the relevant statistical results obtained from the optimization results.

As expected, the GA outperforms other alternatives following the DE method to provide a higher coverage rate (71.64% and 67.20%, respectively). However, both methods spend more time searching for the optimal solution, while PSO shows faster functionality than the other three methods.

- **Case 6:**

The last case allowed all four optimizers to use 1,200 function evaluations to investigate the optimal solution in the WSN coverage problem with 60

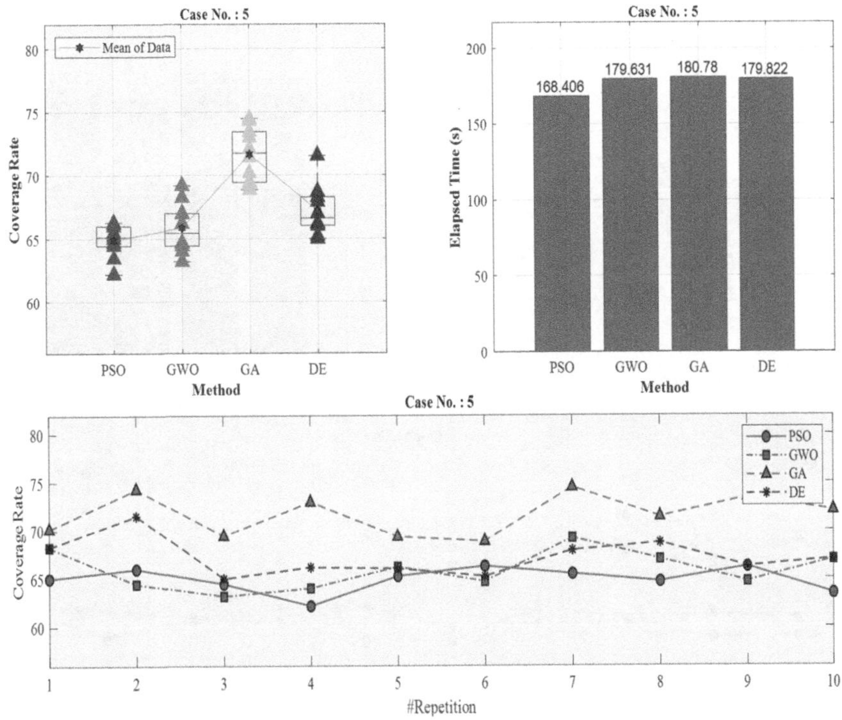

FIGURE 4.6 Box plot, total elapsed time, and representation of obtained fitness function (coverage rate) by different optimizers over 10 individual repetitions for the fifth case in the optimization problem of WSN coverage.

TABLE 4.6 Statistical Outcomes Achieved by Various Optimizers in the Case 5 WSN Coverage Optimization Problem

Method	Ave.	Std.	Max.	Min.	Time Elapsed (s)
PSO	64.89	1.29	66.26	62.16	168.41
GWO	65.85	1.97	69.18	63.18	179.63
GA	71.64	2.11	74.47	68.85	180.78
DE	67.20	1.97	71.53	65.00	179.82

design variables (regarding the coordinates of 30 sensor nodes in a 2D network area). For each optimizer, the size of the initial population is adjusted to 60 and 20 iterations for max number of iterations accordingly. Each optimizer individually repeated 10 times to check the reliability of each method in the current case. All optimization results are recorded. Figure 4.7 illustrates the results of each metaheuristic technique in maximizing the percent coverage rate in the current case.

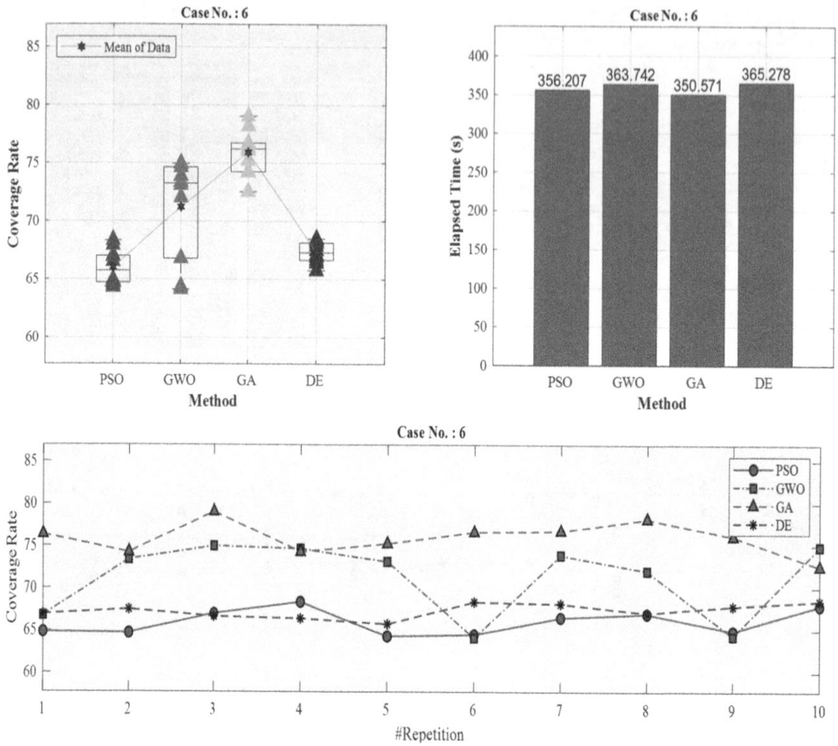

FIGURE 4.7 Box plot, total elapsed time, and representation of obtained fitness function (coverage rate) by different optimizers over 10 individual repetitions for the sixth case in the optimization problem of WSN coverage.

TABLE 4.7 Statistical Outcomes Achieved by Various Optimizers in the Case 6 WSN Coverage Optimization Problem

Method	Ave.	Std.	Max.	Min.	Time Elapsed (s)
PSO	66.00	1.52	68.34	64.31	356.21
GWO	71.22	4.37	74.95	64.14	363.74
GA	75.92	1.94	79.04	72.54	350.57
DE	67.31	0.91	68.49	65.75	365.28

Additionally, the statistical results are provided in Table 4.7.

It is apparent that GA still outperforms other alternatives to provide a higher coverage rate, followed by GWO. Surprisingly, in the current case, by considering the higher number of iterations (20 iterations) compared with the previous case by 10 iterations for optimizers, the GA runs faster with the smallest time elapsed against other methods, including PSO, GWO, and DE.

- **Overall results:**

The plots in Figure 4.8 show the distribution for optimal positioning of 15 sensor nodes with the maximum coverage rate between 10 individual repetitions which are obtained by each method and in each case.

The same plots are provided in Figure 4.9 for 30 sensor nodes.

In addition, the overall overview of all cases, including coverage rate (fitness function) and total computational time elapsed by each optimizer over 10 individual repetitions, is shown in Figure 4.10.

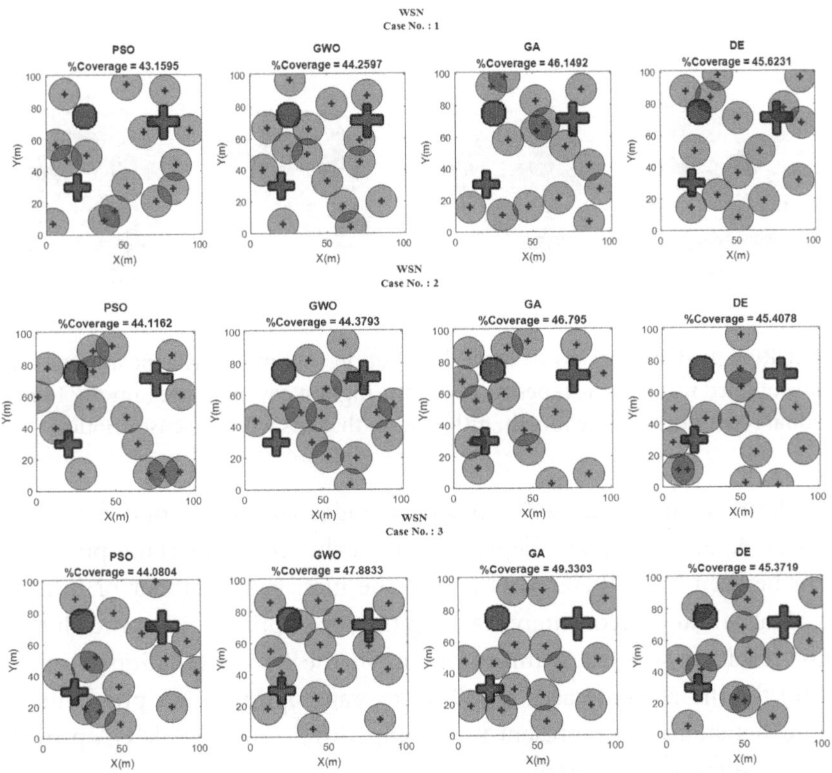

FIGURE 4.8 Comparison of node distribution (optimal sensor positions) using the obtained result in the repetition with the highest coverage rate (among 10 individual repetitions) in the first, second, and third cases with 15 sensor nodes.

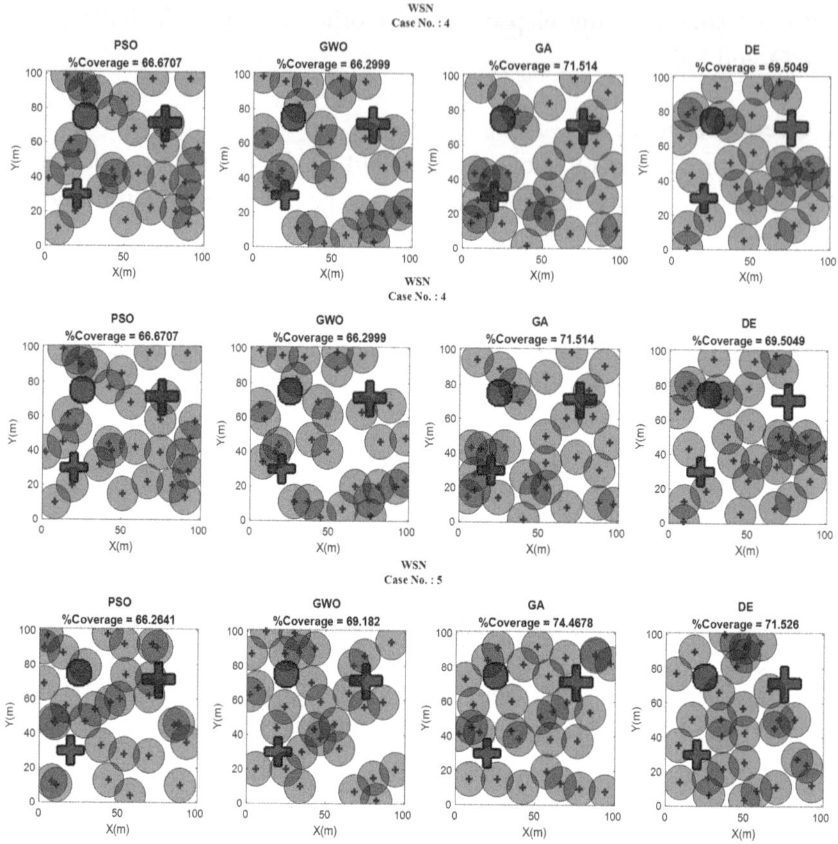

FIGURE 4.9 Comparison of node distribution (optimal sensor positions) using the obtained result in the repetition with the highest coverage rate (among 10 individual repetitions) in the fourth, fifth, and sixth cases with 30 sensor nodes.

The first three cases with 15 sensor nodes and the second three cases (cases 4, 5, and 6) are depicted separately. The important point that can be seen from these plots is that the increasing number of function evaluations and accordingly elapsing time for optimization procedure do not result in significant improvement in the fitness function (coverage rate) for the current cases of WSN coverage optimization problems. All four metaheuristics represent very competitive results, while GA provides much better results overall.

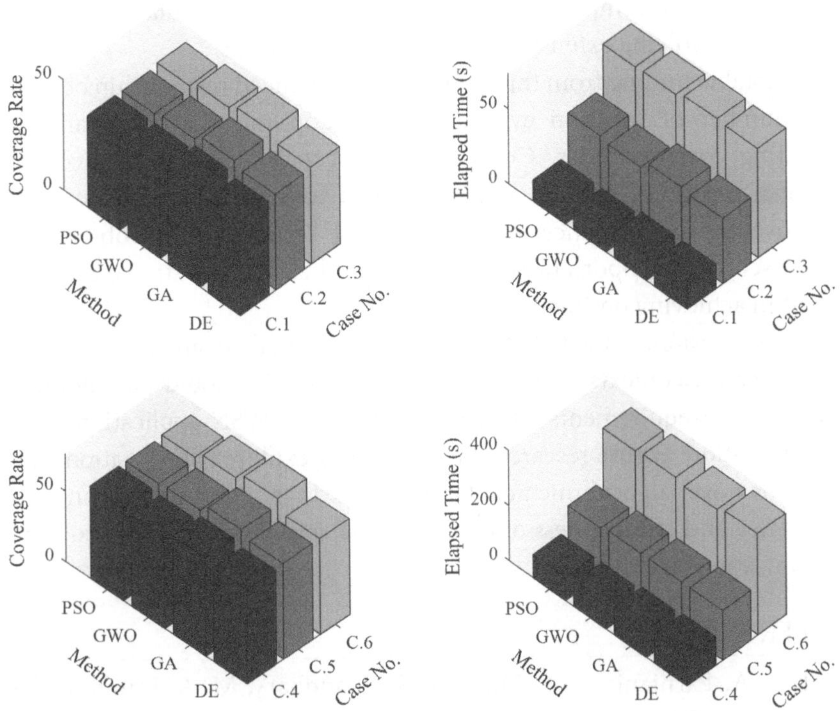

FIGURE 4.10 The overall results including coverage rate (fitness function) and total computational time elapsed by each optimizer over 10 individual repetitions in different cases of WSN coverage optimization problem.

4.6 CONCLUSION

The extensive utilization of metaheuristic algorithms for optimizing WSN coverage remains an area of active exploration within the research community. This chapter contributes to this ongoing discourse through a comprehensive comparative study that systematically evaluates the efficacy of four widely employed metaheuristic techniques: PSO, GWO, GA, and DE. These algorithms were rigorously tested across various configurations of the WSN coverage optimization problem, yielding insights into their respective strengths and limitations.

Throughout the comparative analysis, all four metaheuristics demonstrated competitive and comparable performance metrics. While no single method emerged as definitively superior for all scenarios, GA exhibited a marginally stronger capability in achieving higher coverage rates in certain contexts. However, it was noted that GA tended to operate at a slower pace compared to PSO, GWO, and DE, particularly evident in scenarios involving extensive computational iterations.

A notable finding from this study was the nuanced relationship between the number of function evaluations, total optimization time, and the resulting coverage rates. Contrary to expectations, increasing computational resources did not uniformly lead to significant enhancements in coverage performance across all tested scenarios. This observation underscores the importance of algorithmic efficiency and strategic deployment in achieving optimal WSN coverage.

In conclusion, while GA demonstrated potential advantages in specific optimization contexts, the choice of metaheuristic should be tailored to the specific requirements and constraints of the WSN application under consideration. Future research directions may explore hybridization strategies or novel algorithmic adaptations to further refine and enhance the efficiency and effectiveness of metaheuristic approaches in WSN coverage optimization.

REFERENCES

1. A. Parnianifard, S. Mumtaz, S. Chaudhary, M. A. Imran, and L. Wuttisittikulkij, A data driven approach in less expensive robust transmitting coverage and power optimization, *Scientific Reports*, 12, p. 17725, 2022.
2. X. Qi, Z. Li, C. Chen, and L. Liu, A wireless sensor node deployment scheme based on embedded virtual force resampling particle swarm optimization algorithm, *Applied Intelligence*, 52, (7), pp. 7420–7441, 2022.
3. S. Meguerdichian, F. Koushanfar, M. Potkonjak, and M. B. Srivastava, Coverage problems in wireless ad-hoc sensor networks, in *Proceedings IEEE INFOCOM 2001. Conference on Computer Communications. Twentieth Annual Joint Conference of the IEEE Computer and Communications Society (Cat. No. 01CH37213)*, 2001, 3, pp. 1380–1387.
4. A. Parnianifard et al., Hybrid metamodeling/metaheuristic assisted multi-transmitters placement planning, *Computers, Materials & Continua*, 68, (1), pp. 569–587, 2021.
5. X. Wang, H. Zhang, S. Fan, and H. Gu, Coverage control of sensor networks in IoT based on RPSO, *IEEE Internet of Things Journal*, 5, (5), pp. 3521–3532, 2018.

6. L. Baidar, A. Rahmoun, P. Lorenz, and M. Mihoubi, Whale optimization approach for optimization problem in distributed wireless sensor network, *ACM International Conference Proceeding Series, (May 2020)*, 2019.
7. A. Duttagupta, A. Bishnu, and I. Sengupta, Maximal breach in wireless sensor networks: Geometric characterization and algorithms, in *International Symposium on Algorithms and Experiments for Sensor Systems, Wireless Networks and Distributed Robotics*, 2007, pp. 126–137.
8. A. Gogu, D. Nace, A. Dilo, and N. Mertnia, Optimization problems in wireless sensor networks, in *2011 International Conference on Complex, Intelligent, and Software Intensive Systems*, 2011, pp. 302–309.
9. O. Zorlu and Ö. K. Şahıngöz, Node placement with evolutionary algorithms for maximum coverage of heterogeneous WSNs, in *2017 25th Signal Processing and Communications Applications Conference (SIU)*, 2017, pp. 1–4.
10. A. Jauhri, J. D. Lohn, and D. S. Linden, A comparison of antenna placement algorithms, in *GECCO 2014 - Companion Publication of the 2014 Genetic and Evolutionary Computation Conference*, 2014, pp. 1223–1230.
11. R. V Kulkarni and G. K. Venayagamoorthy, Particle swarm optimization in wireless-sensor networks: A brief survey, *IEEE Transactions on Systems, Man, and Cybernetics, Part C (Applications and Reviews)*, 41, (2), pp. 262–267, 2010.
12. D. Whitley, A genetic algorithm tutorial, *Statistics and Computing*, 4, (2), pp. 65–85, 1994.
13. R. Eberhart and J. Kennedy, Particle swarm optimization, in *Proceedings of the IEEE International Conference on Neural Networks*, 1995, 4, pp. 1942–1948.
14. M. Dorigo and M. Birattari, *Ant Colony Optimization.* Springer, 2010.
15. S. Mirjalili, The ant lion optimizer, *Advances in Engineering Software*, 83, pp. 80–98, 2015.
16. M. Pant, H. Zaheer, L. Garcia-Hernandez, and A. Abraham, Differential evolution: A review of more than two decades of research, *Engineering Applications of Artificial Intelligence*, 90, p. 103479, 2020.

17. S. Mirjalili, S. M. Mirjalili, and A. Lewis, Grey wolf optimizer, *Advances in Engineering Software*, 69, pp. 46–61, 2014.
18. S. Afzal, A Review of Localization Techniques for Wireless Sensor Networks, *Journal of Basic and Applied Scientific Research*, 2, (8), pp. 7795–7801, 2012.
19. S. Kumar, R. Sharma, and E. R. Vans, Localization for wireless sensor networks: A neural network approach, *International Journal of Computer Networks and Communications*, 8, (1), pp. 61–71, 2016.
20. R. Rajakumar, J. Amudhavel, P. Dhavachelvan, and T. Vengattaraman, GWO-LPWSN: Grey wolf optimization algorithm for node localization problem in wireless sensor networks, *Journal of Computer Networks and Communications*, 2017, pp. 1–10, 2017.
21. S. Afzal and H. Beigy, A localization algorithm for large scale mobile wireless sensor networks: A learning approach, *Journal of Supercomputing*, 69, (1), pp. 98–120, 2014.
22. X. Cheng, B. Narahari, R. Simha, M. X. Cheng, and D. Liu, Strong minimum energy topology in wireless sensor networks: NP-completeness and heuristics, *IEEE Transactions on Mobile Computing*, 2, (3), pp. 248–256, 2003.
23. W. Aioffi, G. Mateus, and F. Quintao, Optimization issues and algorithms for wireless sensor networks with mobile sink, in *International Network Optimization Conference*, 2007, 2007, pp. 1–6.
24. B. Yuan, M. Orlowska, and S. Sadiq, On the optimal robot routing problem in wireless sensor networks, *IEEE Transactions on Knowledge and Data Engineering*, 19, (9), pp. 1252–1261, 2007.
25. W. Liu and Z. Fan, Coverage optimization of wireless sensor networks based on chaos particle swarm algorithm, *Journal of Computer Applications*, 2, 2011.
26. N. A. B. Ab Aziz, A. W. Mohemmed, and M. Y. Alias, A wireless sensor network coverage optimization algorithm based on particle swarm optimization and Voronoi diagram, in *2009 International Conference on Networking, Sensing and Control*, 2009, pp. 602–607.
27. H. G. Beyer and B. Sendhoff, Robust optimization – A comprehensive survey, *Computer Methods in Applied Mechanics and Engineering*, 196, (33), pp. 3190–3218, 2007.

28. A. Parnianifard, R. Chancharoen, G. Phanomchoeng, and L. Wuttisittikulkij, A new approach for low-dimensional constrained engineering design optimization using design and analysis of simulation experiments, *International Journal of Computational Intelligence Systems*, 13, (1), pp. 1663–1678, 2020.
29. J. Stork *et al.*, Open issues in surrogate-assisted optimization, in Bartz-Beielstein T., Filipič B., Korošec P., Talbi EG. (eds) *High-Performance Simulation-Based Optimization. Studies in Computational Intelligence*, 833, Springer, Cham, 2020, pp. 225–244.
30. N. Bartoli *et al.*, Adaptive modeling strategy for constrained global optimization with application to aerodynamic wing design, *Aerospace Science and Technology*, 90, (April), pp. 85–102, 2019.
31. A. Parnianifard and A. S. Azfanizam, Metamodel-based robust simulation-optimization assisted optimal design of multiloop integer and fractional-order PID controller, *International Journal of Numerical Modelling: Electronic Networks, Devices and Fields*, 33, (1), p. e2679, 2020.
32. A. Parnianifard, M. Fakhfakh, M. Kotti, A. Zemouche, and L. Wuttisittikulkij, Robust tuning and sensitivity analysis of stochastic integer and fractional-order PID control systems: Application of surrogate-based robust simulation-optimization, *International Journal of Numerical Modelling: Electronic Networks, Devices and Fields*, 34, (2), p. e2835, 2020.
33. A. Parnianifard, A. S. Azfanizam, M. K. A. Ariffin, and M. I. S. Ismail, Kriging-assisted robust black-box simulation optimization in direct speed control of DC motor under uncertainty, *IEEE Transactions on Magnetics*, 54, (7), pp. 1–10, 2018.
34. F. A. C. Viana, T. W. Simpson, V. Balabanov, and V. Toropov, Metamodeling in multidisciplinary design optimization: How far have we really come? *AIAA Journal*, 52, (4), pp. 670–690, 2014.
35. A. Mohammad Nezhad and H. Mahlooji, An artificial neural network meta-model for constrained simulation optimization, *Journal of the Operational Research Society*, 65, (8), pp. 1232–1244, 2013.
36. A. Parnianifard, S. Chaudhary, S. Mumtaz, L. Wuttisittikulkij, and M. A. Imran, Expedited surrogate-based quantification of engineering tolerances using a modified polynomial regression, *Structural and Multidisciplinary Optimization*, 66, (3), p. 61, 2023.

37. E. Bonabeau, D. de R. D. F. Marco, M. Dorigo, and G. Theraulaz, *Swarm Intelligence: From Natural to Artificial Systems*, (1). Oxford University Press, 1999.
38. A. Parnianifard, A. S. Azfanizam, M. K. A. Ariffin, and M. I. S. Ismail, An overview on robust design hybrid metamodeling: Advanced methodology in process optimization under uncertainty, *International Journal of Industrial Engineering Computations*, 9, (1), pp. 1–32, 2018.
39. A. Parnianifard, A. Azfanizam, M. Ariffin, M. Ismail, and N. Ebrahim, Recent developments in metamodel based robust black-box simulation optimization: An overview, *Decision Science Letters*, 8, (1), pp. 17–44, 2019.
40. A. Parnianifard, V. Rezaie, S. Chaudhary, M. A. Imran, and L. Wuttisittikulkij, New adaptive surrogate-based approach combined swarm optimizer assisted less tuning cost of dynamic production- inventory control system, *IEEE Access*, 9, pp. 144054–144066, 2021.
41. A. Parnianifard, A. Zemouche, R. Chancharoen, M. A. Imran, and L. Wuttisittikulkij, Robust optimal design of FOPID controller for five bar linkage robot in a cyber-physical system: A new simulation-optimization approach, *PLoS One*, 15, (11), p. e0242613, Nov. 2020.
42. A. Parnianifard, A. S. Azfanizam, M. K. A. Ariffin, and M. I. S. Ismail, Crossing weighted uncertainty scenarios assisted distribution-free metamodel-based robust simulation optimization, *Engineering with Computers*, 36, (1), pp. 139–150, 2019.
43. A. Parnianifard, A. S. Azfanizam, M. K. A. Ariffin, and M. I. S. Ismail, Comparative study of metamodeling and sampling design for expensive and semi-expensive simulation models under uncertainty, *Simulation*, 96, (1), pp. 89–110, May 2019.
44. S. Wang, X. Yang, X. Wang, and Z. Qian, A virtual force algorithm-Lévy-embedded grey wolf optimization algorithm for wireless sensor network coverage optimization, *Sensors (Switzerland)*, 19, (12), p. 2735, 2019.
45. H. Faris, I. Aljarah, M. A. Al-Betar, and S. Mirjalili, Grey wolf optimizer: A review of recent variants and applications, *Neural Computing and Applications*, 30, (2), pp. 413–435, 2018.
46. X. Wang, H. Gu, Y. Liu, and H. Zhang, A two-stage RPSO-ACS based protocol: A new method for sensor network clustering

and routing in mobile computing, *IEEE Access*, 7, pp. 113141–113150, 2019.
47. D. Wang, H. Wang, X. Ban, X. Qian, and J. Ni, An adaptive, discrete space oriented wolf pack optimization algorithm for a movable wireless sensor network, *Sensors*, 19, (19), p. 4320, 2019.
48. A. Adulyasas, Z. Sun, and N. Wang, Connected coverage optimization for sensor scheduling in wireless sensor networks, *IEEE Sensors Journal*, 15, (7), pp. 3877–3892, 2015.
49. S. Wang, X. Yang, X. Wang, and Z. Qian, A virtual force algorithm-Lévy-embedded grey wolf optimization algorithm for wireless sensor network coverage optimization, *Sensors*, 19, (12), p. 2735, 2019.
50. R. Song, Z. Xu, and Y. Liu, Wireless sensor network coverage optimization based on fruit fly algorithm, *International Journal of Online Engineering*, 14, (6), pp. 58–70, 2018.
51. W.-H. Liao, Y. Kao, and R.-T. Wu, Ant colony optimization based sensor deployment protocol for wireless sensor networks, *Expert Systems with Applications*, 38, (6), pp. 6599–6605, 2011.
52. A. A. A. Ari *et al.*, Bacterial foraging optimization scheme for mobile sensing in wireless sensor networks, *International Journal of Wireless Information Networks*, 24, (3), pp. 254–267, 2017.
53. I. Vilovic, N. Burum, and Z. Sipus, Ant colony approach in optimization of base station position, in *2009 3rd European Conference on Antennas and Propagation*, 2009, pp. 2882–2886.
54. C. K. Ting, C. N. Lee, H. C. Chang, and J. S. Wu, Wireless heterogeneous transmitter placement using multi-objective variable-length genetic algorithm, *IEEE Transactions on Systems, Man, and Cybernetics, Part B (Cybernetics)*, 39, (4), pp. 945–958, 2009.
55. R. L. Haupt and D. H. Werner, *Genetic Algorithms in Electromagnetics*. John Wiley & Sons, 2007.
56. Z. Yangyang, J. Chunlin, Y. Ping, L. Manlin, W. Chaojin, and W. Guangxing, Particle swarm optimization for base station placement in mobile communication, in *IEEE International Conference on Networking, Sensing and Control, 2004*, 2004, 1, pp. 428–432.
57. Z. Miao, X. Yuan, F. Zhou, X. Qiu, Y. Song, and K. Chen, Grey wolf optimizer with an enhanced hierarchy and its application to the wireless sensor network coverage optimization problem, *Applied Soft Computing*, 96, (August), p. 106602, 2020.

58. J. J. Grefenstette, Genetic algorithms and machine learning, in *Proceedings of the Sixth Annual Conference on Computational Learning Theory*, 1993, pp. 3–4.
59. S. Mirjalili and S. Mirjalili, Genetic algorithm, in Mirjalili S. (ed) *Evolutionary Algorithms and Neural Networks: Theory and Applications*, Springer, Cham, pp. 43–55, 2019.
60. R. Storn, On the usage of differential evolution for function optimization, in *Proceedings of North American Fuzzy Information Processing*, 1996, pp. 519–523.
61. Bilal, M. Pant, H. Zaheer, L. Garcia-Hernandez, and A. Abraham, Differential evolution: A review of more than two decades of research, *Engineering Applications of Artificial Intelligence*, 90, (January), p. 103479, 2020.
62. K. V Price, Differential evolution, in Zelinka I., Snášel V., Abraham A. (eds) *Handbook of Optimization: From Classical to Modern Approach*, Springer, Berlin, pp. 187–214, 2013.

CHAPTER 5

Fundamentals of Machine Learning in Communication Systems

Shivani Sharma and Avani Vyas

5.1 INTRODUCTION

The demands for security, availability, and performance of new applications, services, and devices are escalating at a pace beyond initial expectations. Real-time responsiveness is crucial in application domains such as e-health, traffic management, and industry operations, necessitating communication networks to autonomously make real-time decisions. Achieving this level of autonomy entails networks that can react to and learn from their environment, effectively controlling themselves without human intervention. However, conventional communication networks have historically followed a different trajectory, relying on manual human responses to factors like traffic fluctuations, updates in network functions, security incidents, and faults. This reliance on human-machine interactions has resulted in network downtime, increased security vulnerabilities, and numerous challenges within current communication networks.

The necessity for human interaction or manual configuration presents a significant obstacle for networks to leverage their past experiences to adapt to changing requirements. The overarching concept involves predicting the future behavior of services, network segments, users, or user equipment

DOI: 10.1201/9781003472506-5

(UE) and dynamically tuning the network based on this predictive information. For example, predicting the movement trajectory of a user can be accomplished through straightforward mechanisms like measuring signal strength between consecutive base stations to minimize handover latency. Furthermore, the evolving landscape of services and networks, such as vehicular communication and the internet of things (IoT) in the fifth generation (5G), underscores the need for the introduction of intelligent network operations. Intelligence in communication networks is characterized by an autonomous pursuit of goals without human intervention, especially in the face of uncertainty arising from a lack of information. The system must actively react to and adapt to changes in its environment, learning from experience through information gathered during its operation.

Machine learning (ML) systems can be broadly classified into several categories, including supervised learning, unsupervised learning, semi-supervised learning, and reinforcement learning. In supervised learning, the system predicts output data based on labeled training data, while unsupervised learning focuses on identifying patterns and associations within input data. Semi-supervised learning involves training with a combination of labeled and unlabeled input data, aiming to either label the unannotated data or predict outputs for new input data. In reinforcement learning, the system learns to take actions that maximize cumulative rewards.

These ML techniques empower systems with intelligence, enhancing their performance based on available information. In the context of communication networks, integrating ML disciplines, technologies, and methods becomes imperative. This integration helps mitigate risks associated with human control, enabling networks to autonomously adapt, self-control, and respond to dynamic conditions.

ML can also contribute to improving the performance of cognitive radios in spectrum sharing and optimizing resources in heterogeneous access networks. Emerging technologies like software-defined networking (SDN), network function virtualization (NFV), and multi-access edge computing (MEC) offer dynamic network operations but come with challenges, such as potential bottlenecks in centralized control platforms. Investigating mechanisms that enable networks to learn from their environment, understand challenges, and dynamically configure themselves is crucial, and ML stands out as a promising candidate for providing such capabilities in communication networks.

5.2 ML FOR PHYSICAL LAYER

The physical layer, often denoted as Layer 1, represents the foundational level within a communication system. It is responsible for overseeing the efficient transmission, receiver processing, and precise modeling of channels to guarantee the dependable transmission of data across physical channels. The development of communication signal processing algorithms for the physical layer typically involves analytical design techniques, utilizing mathematical optimization, statistics, and information theory. In this section, certain key domains for ML applications in the physical layer are discussed in detail.

5.2.1 Application of Neural Networks for Digital Pre-distortion

Non-linear distortion poses significant challenges for digital systems, primarily stemming from high-power amplifiers (HPAs) in digital radio systems. While linear class A amplifiers and operating the HPA away from saturation are conventional solutions, they result in expensive and inefficient systems. A more efficient approach involves employing a non-linear filter at the transmitter, creating the inverse of an HPA response, known as digital pre-distortion (DPD). In the depicted typical transmitter setup, the baseband input stream a(KT0) undergoes pre-distortion, yielding the signal b(KT0), which then passes through modulator and demodulator pulse shaping filters with impulse responses gT(t) and gR(t). The impact of the HPA is assessed by comparing the demodulator output c(KT0) with the pre-distortion filter output b(KT0). The pre-distortion filter's weight adjusts based on the disparity between c(KT0) and b(KT0). However, the accuracy of the pre-distortion filter hinges on the precision of the HPA model. Traditional memory polynomial models may falter due to the HPA's non-linear continuous function response, especially when accommodating various signal types. Artificial neural networks (ANNs) emerge as an efficient tool for implementing non-linear mappings. With sufficient neurons, feed-forward ANNs act as universal approximators for arbitrary continuous functions, making them a natural choice for pre-distortion, trainable to generalize HPA models across different carrier types.

In reference [1], an ANN-based pre-distortion technique for satellite communications was introduced by the author. The proposed DPD method comprises two distinct ANNs. The initial ANN is employed for modeling the transfer function of the traveling wave tube (TWT) amplifier, while the second ANN is utilized for the inverse of the transfer function. Both ANNs are structured with two layers, each consisting of $N = 10$ neurons. In contrast to using a known normalized TWT, the authors assumed a non-normalized solid-state HPA with unspecified parameters and intrinsic adaptive behavior in a different study [2]. Each ANN in this case is composed of a hidden layer with $N = 10$ neurons and an output layer with a single neuron.

5.2.2 Learning to Decode

Channel coding is a strategy employed to manage errors in data communication over noisy channels. The utilization of ANNs in the realm of channel coding and decoding was first introduced by Bruck and Blaum in 1989 [3]. Their pioneering work illustrated the construction of an ANN, associated with an error-correction code, wherein each local maximum is representative of a codeword, and vice versa. Subsequent research, detailed in Zeng et al. [4], demonstrated the practical application of ANN in the decoding of error-correction codes. Furthermore, Caid and Means [5] delved into the implementation of ANN for channel decoders. The proposed ANN-based decoders exhibited superior performance compared to traditional counterparts when faced with scenarios where assumptions, such as additive white Gaussian noise (AWGN) or binary symmetric channel (BSC), were violated, such as in jamming environments. A departure from conventional methods was proposed by using an ANN for Hamming codes, allocating each output node to a codeword instead of individual bits. The authors explored potential solutions for Hamming codes using two distinct ANNs: the counter-propagation neural network and back-propagation.

An alternative approach is presented in Liang et al. [6], which proposes an iterative belief propagation scheme employing a convolutional neural network (CNN) architecture. Here, a conventional belief propagation decoder estimates coded bits, and a CNN is subsequently employed to rectify estimation errors. In Lyu et al. [7], a comparative analysis is conducted for channel decoding, evaluating the performance of multilayer perceptron (MLP), CNN, and recurrent neural network (RNN). The findings reveal that RNN exhibits superior decoding performance, albeit with increased complexity. Additionally, the study notes that the codeword length significantly influences the adaptability of ANNs, coining the term "saturation length" to denote the point at which learning becomes restricted.

5.2.3 Neurobeamformer

Beamforming is a signal processing technique crucial for directional signal transmission and reception. It revolves around adjusting the phase angles of an antenna array to create constructive interference for signals at a specific angle, thereby focusing the signal in the desired direction. The calculation of phase angles typically takes place in the digital domain, forming an integral part of the physical layer. Unlike pre-equalization or pre-coding in MIMO systems, beamformers involve steering a beam toward a specific direction. The historical evolution of ANN-based beamformers traces back to the 1980s. Figure 5.1 depicts a typical configuration of a neurobeamformer.

ANN is employed to predict the beamforming weights. This neurobeamformer utilizes the ANN to fine-tune the angles of the phased arrays, ensuring that transmitted signals from various antennas converge precisely in the target user's direction. The concept of radial basis function artificial neural network (RBF ANN) is expanded upon by Southall et al. [8]. In this work, the authors employ an adaptive RBF ANN trained through a gradient descent algorithm and a linear algebra-based network

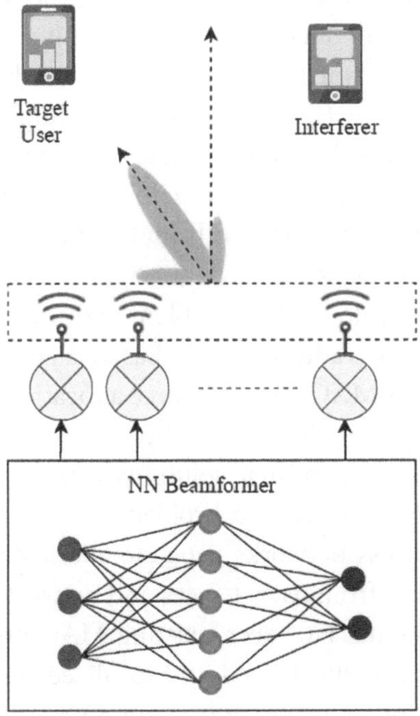

FIGURE 5.1 A typical setup of a neurobeamformer where an ANN is used to predict the beamforming weights.

using an LMS error solution. While the ANN beamformers demonstrate high performance, challenges arise with data subjected to severe near-field scattering conditions.

Lovato and Gong [9], introduced a novel ANN structure for implementing antenna array beamforming. This ANN comprises two hidden layers, with the first layer divided into sublayers equal to the number of inputs. These sublayers are fully connected to the second hidden layer. The ANN is trained using data from minimum variance distortionless response (MVDR) beamforming. The proposed ANN structure surpasses conventional ANN structures in terms of beamforming performance.

5.3 ML FOR MEDIUM ACCESS CONTROL LAYER

In wireless networks, the spectrum is a limited resource, necessitating a robust medium access control (MAC) protocol that facilitates channel access for multiple users. The objective is to maximize spectrum utilization and ensure quality of service (QoS). As the number of connected devices continues to rise, expanding the optimization domain with an increasing number of parameters and stringent QoS requirements, MAC protocols are expected to face unprecedented complexity. The MAC protocol's intricacy depends on factors such as network architecture, communication model, and duplex mechanisms (e.g., time division duplexing (TDD), frequency division duplexing (FDD), and full duplexing). The constant evolution of wireless networks, introducing new technological components like flexible duplexing, adaptive frame numerologies, and the convergence of heterogeneous wireless networks with multi-interface radio devices, further complicates MAC tasks.

Viewed as a large-scale control problem with diverse QoS constraints, optimizing MAC with traditional rule-based algorithms is suboptimal. Current algorithms for MAC radio resource allocation often adopt an optimization approach that makes assumptions to simplify the non-convex problem, resulting in suboptimal solutions due to the complexity that hinders solving the full non-convex problem.

Furthermore, wireless networks generate abundant data through continuous measurements from user devices and network entities in the form of radio and system data. However, existing MAC protocols underutilize this data, considering it short-lived and localized due to aging and user mobility. The evolution of machine learning (ML) and wireless networks presents an opportunity to leverage such data in multiple dimensions, creating data-driven wireless networks that are more robust and autonomous in dynamic environments. This section provides a comprehensive overview of potential research directions and associated challenges arising from the integration of ML algorithms into MAC protocol design for future wireless networks.

5.3.1 Interference Prediction

The transmission of signals by multiple devices in the same frequency within the shared medium of air disrupts signal reception, a phenomenon known as "interference." Despite advancements in modeling interference dynamics in wireless networks using approaches like interference alignment [10], researchers have struggled to develop algorithms that fully leverage interference knowledge to enhance the QoS of communication.

Interference alignment's fundamental concept involves coordinating multiple transmitters to align mutual interference at the receiver. However, ongoing efforts explore the integration of machine learning, such as using an autoregressive (AR) model, where the next state is predicted based on past states, to enhance the performance of interference alignment algorithms [11]. A key challenge in utilizing interference information lies in its randomness and short-time validity due to mobility. Additionally, identifying the exact source of interference among multiple sources, coupled with the random nature of interferers at each time instance, further compounds the complexity of the problem.

5.3.2 User Mobility and Association

Numerous researchers have delved into employing machine learning (ML) techniques for localization, aiming to enhance algorithm accuracy by training them with phase information and estimating the angle of arrival through ML using phase fingerprinting [12–15]. In Ezpeleta et al. [12], an indoor positioning algorithm based on KNN utilizes historical data for estimating the mobile user's position. Similarly, the algorithm presents a positioning estimation algorithm based on deep neural network (DNN). This algorithm employs a stacked denoising autoencoder (SDA) and a hidden-Markov model (HMM) to minimize the feature set, smoothing the original location without compromising information content.

Zhang et al. [13] combine support vector machine (SVM) and ANN to estimate user position based on the received signal strength indicator (RSSI). However, the focus of this work is on boundary-level localization rather than the precise user position. In Berruet et al. [15], a CNN-based fingerprinting technique for indoor localization based on channel frequency response (CFR) is introduced. Similarly, another scheme has been proposed, i.e. deep long short-term memory (LF-DLSTM) approach for indoor localization based on RSSI, reducing noise effects to improve positioning accuracy.

However, a prevalent issue in many proposed positioning algorithms, whether based on ML techniques or conventional methods like OTDOA, UTDOA, RSSI, and Cell-ID, is the reliance on single value estimates (SVEs) such as phase information, angle of arrival, or RSSI. Localization estimates heavily depend on the quality of these SVEs, which degrade or exhibit high randomness in the wireless environment due to factors like multipath and non-line-of-sight (NLOS). To fully leverage the potential of learning techniques in terms of user mobility and localization, an avenue worth exploring is cooperative positioning techniques through data fusion of 5G with other sources like cameras, device-to-device (D2D) links, GPS, etc. Additionally, optimizing communication and positioning targets jointly, as they often overlap, represents a promising direction.

5.3.3 Radio Resource Scheduling

Machine learning (ML) algorithms offer innovative solutions for addressing asymmetric traffic accommodation in wireless networks, particularly by enhancing the flexible duplexing capabilities of 5G networks—a promising technique for optimizing resources based on traffic demand in both downlink and uplink [16]. Flexible duplexing aims to achieve adaptive resource allocation considering the asymmetric uplink and downlink traffic, optimizing in the time-frequency domain and blurring the distinction between TDD and FDD, or even eliminating it entirely. In Malik et al. [17], the author introduces radio resource allocation based on flexible duplexing to mitigate inter-cell interference.

To fully harness the potential of flexible duplexing and adaptively optimize resources, predicting or learning the traffic behavior of the network becomes crucial. Numerous researchers have explored traffic flow prediction using linear models such as autoregressive (AR), autoregressive moving average (ARMA), and moving average (MA) [18,19], achieving success in small-scale sparse networks. However, considering the evolving nature of wireless networks, many non-linear factors, such as multi-dependency and abruptness, significantly impact performance. Therefore, recent efforts have focused on machine learning-based models involving ANNs, SVM, least-square support vector machines (LSSVM), long short-term memory (LSTM), etc. [20]. These models learn from historical traffic data to understand randomness and improve traffic flow prediction. In Zhao et al. [20], a 2D-LSTM model enables self-optimization and an intelligent response for fault tolerance, showcasing the ability of ML techniques to provide self-X features essential for HetNets.

Moreover, future wireless systems like NB-IoT and LTE-M are expected to adopt symmetric frequency allocation between downlink and uplink due to half-duplex frequency division duplex (HD-FDD). However, such symmetric allocation may lead to underutilization of spectrum resources, given the asymmetric nature of traffic. ML techniques can intelligently address such challenges, especially in IoT networks, where uplink data traffic is anticipated to be higher than downlink, potentially avoiding resource wastage. Similarly, in broadband communication, where downlink traffic prevails, ML techniques can offer intelligent solutions to seamlessly integrate environmental parameters and enable proactive MAC functionalities, departing from traditional reactive approaches.

5.4 ML FOR NETWORK LAYER

In the realm of network research, the network layer assumes a pivotal role in orchestrating the end-to-end delivery of packets through the calculation of routes across the network [21]. As the landscape of communication entities grows increasingly complex and diverse, the imperative arises for the network to autonomously configure routing nodes with minimal human intervention. This configuration involves adjusting the behavior of network devices, such as routers and switches, to facilitate the seamless delivery of traffic from various applications. However, this heightened complexity introduces a set of challenges.

Traditional network traffic control systems often rely on rigid, predefined policies, rendering the network less adaptable to dynamic environments and susceptible to stagnation. This predicament becomes particularly pronounced in wireless networks, where abrupt shifts in traffic patterns and volumes are commonplace. Consequently, there is a pressing need for novel mechanisms to ensure resilient and optimal end-to-end packet delivery. Within the realm of network research, machine learning (ML) has emerged as one of the most effective mechanisms for addressing and surmounting these challenges in network configuration and traffic control.

5.4.1 Network Traffic Control

Effective control of network traffic is essential for optimizing overall network performance, whether in fixed networks [22], wireless networks [23,24], or virtual networks [25]. Network traffic engineering plays a crucial role in reducing congestion and enhancing utilization by distributing the load across multiple paths. By doing so, it enables communication networks to deliver data optimally between two points, even in the event of link or router failures. However, the prevalent approach of independently performing failure recovery and traffic engineering in today's networks often leads to complex and compromised traffic control systems [26]. The evolving landscape of novel services and devices necessitates automated network traffic control systems, prompting the application of ML-based approaches for efficiency.

Deep learning (DL) has been explored for intelligent traffic control, as highlighted by Fadlullah et al. [27], where the authors provide an overview of state-of-the-art DL architectures and algorithms relevant to network traffic control systems. The core function of a network traffic control system is to efficiently route packets between two entities. Traditional routing protocols, being non-intelligent, cannot learn from past events for autonomous decision-making [28]. Consequently, there is a shift toward replacing conventional network protocols, reliant on algorithmic insights from human experts, with data-driven, predictive, or ML-based approaches. Traffic steering is another avenue where ML techniques can be employed to avoid congested paths in the network or maximize QoS among existing alternative paths, as exemplified by Adachi et al. [29] through Q-learning. Figure 5.2 presents a simple network traffic control system using ML, where traffic patterns are used to train the system for optimal routing paths.

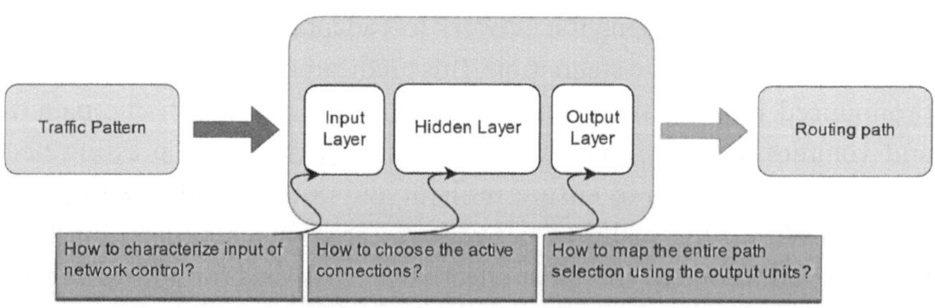

FIGURE 5.2 General learning approaches for network traffic control.

The subsequent sub-sections delve into ML-aided routing protocols tailored for diverse network architectures and networking technologies.

5.4.2 Routing

The application of ML for routing has been explored by Valadarsky et al. [30]. The primary focus of this work is to assess whether ML can be effectively employed to automatically generate optimal routing configurations. The study, framed within the context of intra-domain traffic engineering [31], serves as a case study. It delves into ML-guided routing optimization within a single, self-administered network, with specific attention given to (i) framing routing as an ML problem, and (ii) identifying suitable representations for input and output within the learning domains. The authors conclude that data-driven routing, or ML-based routing, holds significant potential for enhancing network performance but can also pose notable challenges if not applied judiciously. These challenges relate to network overhead and delay, particularly when proper techniques and learning parameters are not meticulously selected.

The effectiveness of ML in improving the quality of transmission in optical networks is demonstrated by Salani et al. [32]. The proposed integrated approach for routing and spectrum assignment leverages probabilistic outputs of ML-based quality of transmission. This approach results in saving up to 30% of the spectrum and a reduction in the number of risky lightpaths—those exceeding a certain threshold of bit error rate.

In the domain of traffic prediction, Musumeci et al. [33] explore supervised learning mechanisms utilizing historical data and user profiles. These mechanisms prove effective in predicting future traffic requirements and the corresponding resource needs.

5.4.3 Routing in Dynamic (AD-HOC) Networks

In Sharma et al. [34], an ML-based protocol for efficient routing in opportunistic networks is proposed. This routing protocol enhances the performance of the PROPHET+ protocol [35] by integrating ML techniques to self-train based on various factors, including hop count, buffer capacity, node energy, node moving speed, node popularity, and the number of successful deliveries. The ML algorithm utilizes past routing data to assess a node's capability to deliver messages to intended destinations. Two ML models, ANN and decision trees (DTs), are employed to calculate the probability of correct message delivery. The proposed protocol enhances delivery probability while minimizing average latency, overhead, and buffer size requirements.

Similarly, in Zhi-yuan et al. [36], the wavelet neural network prediction model is utilized for multipath routing in wireless mesh networks, demonstrating improved adaptability and robustness to link failures and congestion. However, conventional ML techniques, such as ANNs, exhibit limitations in terms of scalability and computation efficiency when considered for routing, as discussed in Kato et al. [37]. Therefore, incorporating additional resource constraints through ML in routing on the data plane or routing devices could exacerbate existing congestion and latency challenges. It is crucial to evaluate the cost versus benefits before adopting ML-based techniques for routing.

5.5 ML FOR EDGE COMPUTING

The prevailing architecture for mobile and IoT services consists of a three-tier structure, incorporating local user interfacing, sensor data collection, and actuation at the edge layer, while service and application logic are centralized in data centers positioned between the cloud and devices. This three-tier model offers advantages, such as resource efficiency at the local level and global service availability. However, the edge paradigm faces significant challenges stemming from two prominent trends: the rapid proliferation of sensor and user-generated data, and the real-time demands posed by modern applications, including augmented and virtual reality [38].

The current centralized cloud architecture has demonstrated shortcomings in meeting the performance and scalability requirements associated with such data-intensive and real-time applications and services [38,39]. This realization has sparked a growing interest in establishing an intelligent edge positioned between end-user/IoT devices and the cloud architecture. This intelligent edge is envisioned to address the evolving needs of applications and services, especially in the context of burgeoning data volumes and the heightened demands for real-time processing and responsiveness. ML plays a dual role in optimizing the operation of edge and fog systems and providing services for mobile applications. At the fog level, ML functions as a tool for optimizing the placement of various application functionalities, making decisions on how to distribute computational tasks across local, edge, and cloud/data center levels. Examples of fog-level optimization include determining the partitioning of computational tasks. On the edge level, ML is employed for tasks such as predicting changes

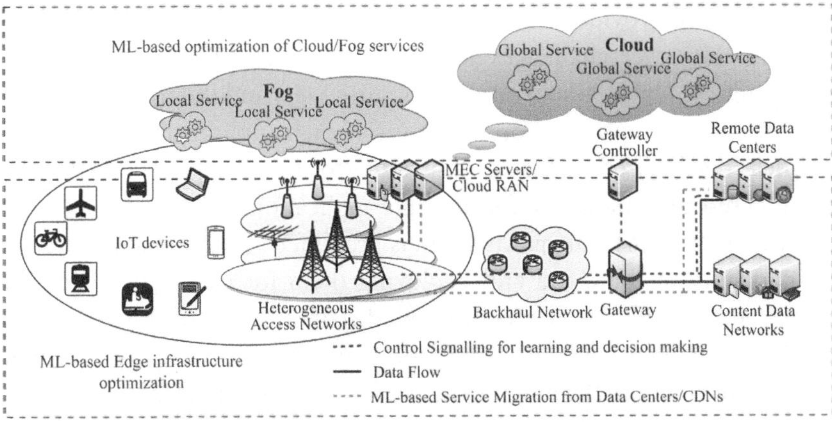

FIGURE 5.3 ML and Edge for service and infrastructure optimization.

in workload, orchestrating and initiating edge computing resources, and making decisions on application offloading, considering factors like workload on edge nodes and user mobility across the edge platform.

Figure 5.3 visually represents the twofold role of ML in optimizing the operation of edge/fog systems and delivering services for mobile applications.

In its second role, ML takes input from the underlying edge/fog platform to enhance its operation. Training ML at the edge over wireless networks, while considering factors like latency and reliability, presents new research directions. ML optimization, based on behavioral and workload information from the edge/fog platform, offers advantages such as the ability to predict changes in workload concerning spatial and temporal constraints and user mobility. This enables real-time orchestration of edge resources and task offloading based on dynamic conditions.

5.5.1 ML-Based Optimization of Cloud/Fog Services

In Aryal and Altmann [40], an intelligent resource allocation algorithm leveraging genetic algorithms is introduced for the optimal distribution of edge and cloud resources. Through an iterative process, the proposed algorithm aims to provide an optimal solution by aligning cloud provider resources with specific application requirements. The evaluations suggest that the algorithm is effective in achieving a Pareto-optimal match of cloud provider resources, and it can be customized to meet the specific needs of applications. The authors highlight future directions, emphasizing the importance of provider resource ranking and predicting application user behavior. However, challenges lie in the optimized utilization of resources with multiple objectives for application providers and the federation operator. The unique optimization objectives of each application, coupled with the heterogeneity of resources, contribute to the complexity of achieving optimal resource allocation for multiple applications.

In Calo et al. [41], an architecture is presented for composing analytics functionality in the cloud and deploying them for use at the edge. This architecture facilitates the controlled distribution of data stream processing, with intensive yet well-structured processing occurring at the edge. The authors identify the main benefits of edge computing (EC) as reducing latency, minimizing data movement, and preserving local data privacy. The functionality split between observer and controller components is explored, and the supporting system tools required for efficiently composing distributed collaborative analysis applications for streaming data are identified. The overarching goal is to enable support for collaborative autonomous systems, primarily through the application of generative policy models.

5.5.2 ML-Based Optimization of the Edge Infrastructure

Numerous proposals and approaches address tasks and resource optimization in MEC platforms and MEC-based networks. In Wang et al. [42], a multi-stack reinforcement learning algorithm is introduced for MEC to optimize tasks or resource allocations for users. The proposed algorithm enhances convergence speed and learning efficiency. Rodrigues et al. [43] delve into efficiently utilizing MEC resources amidst a growing number of complex devices and services with an increasing number of configuration parameters, leveraging ML. The work in Rodrigues et al. [43] also identifies computation and communication challenges related to MEC that can be addressed by various ML solutions.

Wu et al. [44] propose and evaluate collaboration between edge and centralized cloud resources using ML, while Nduwayezu et al. [45] employ deep reinforcement learning (DRL) to enhance online computing offloading for non-orthogonal multiple access.

Several proposals focus on specific technologies within MEC through ML interventions. For instance, Cao et al. [46] and Cui et al. [47] use multi-agent RL for cooperative caching and task offloading in MEC, respectively. Few works contributed to DRL-based energy-efficient task offloading for machine-type communication in the edge. Furthermore, some address minimizing per-bit energy consumption in 5G MECs using DNNs, improving user energy efficiency by developing a digital twin of the real system and training the DL system within it. Similarly, Cui et al. [48] propose stochastic online learning for MEC, aiming to minimize the time-averaged operational costs of MECs. Subramanya et al. [49] focus on ML-based service function chain placement and scaling in MEC, while still there is the need for AI or ML-based offloading in MEC and review state-of-the-art research in this direction.

5.5.3 Edge-Based ML Optimization

Federated learning (FL) emerges as a compelling approach for edge ML training architecture, as discussed in Chen et al. [50] and Shlezinger et al. [51]. FL allows computing nodes to collaboratively learn a shared prediction model while keeping training data decentralized on individual devices. In McMahan et al. [52], a method for FL of deep networks based on iterative model averaging is presented, demonstrating robustness to unbalanced and non-identically distributed (non-IID) data distributions. Communication costs are recognized as a primary constraint, but the experiments reveal that FL can be made feasible in constrained environments, training high-quality models with relatively few communication rounds compared to synchronized stochastic gradient descent. The authors emphasize the practical privacy benefits of FL, including stronger guarantees via differential privacy, secure multi-party computation, and their combination. They view FL as a promising avenue for future research.

Chen et al. [50] introduce a framework for implementing FL algorithms in wireless networks, optimizing collaboration between cellular base stations and wireless users to address challenges inherent in the wireless nature of the network. Convergence time optimization in FL is explored in Chen et al. [53] and [54], with a focus on minimizing convergence time while enhancing FL performance in resource-constrained cellular wireless environments.

In Konecný et al. [55], FL is studied for optimizing ML prediction models using local training data. The authors propose a new algorithm for sparse convex problems, demonstrating the possibility of designing efficient algorithms utilizing local training data without leaving the source node at any phase. The work sets the stage for future research directions, including the development of a fully asynchronous version of the proposed algorithm and a deeper theoretical understanding.

Additionally, Lim et al. [56] present a survey on using FL in MEC platforms. The article explores the background of FL in MEC, implementation challenges, and solutions and outlines future research directions in this evolving field.

5.6 FUTURE RESEARCH DIRECTIONS

Here, we discuss a few important research areas that require significant contributions for resolving some real problems the industry is still facing.

1. **Future of ML in Physical Layer:** The application of ML techniques holds significant promise for enhancing the physical layer. Nevertheless, several inherent challenges need to be addressed. One such challenge involves the necessity of standardized benchmarks and datasets, which play a crucial role in assessing the performance of different ML algorithms. Despite this, there is currently a lack of comparable benchmarks or shared datasets specific to the physical layer. Further exploration is required to apply ML disciplines, particularly deep learning (DL), to the physical layer, addressing various perspectives. The efficacy of DL-based modulation recognition methods underscores their utility in feature extraction and recognition within this domain. Furthermore, integrating intelligence at the physical layer with higher layers of the network has the potential to create a versatile and intelligent overall network. The exploration of cross-layer intelligence sharing through ML techniques remains underexplored, presenting an intriguing avenue for future research.
2. **Future of ML in Medium Access Control Layer:** While advancements have been achieved in proactive MAC algorithms by integrating sophisticated learning techniques, there remains a pressing need to explore and formulate more robust and stringent solutions to address challenges such as interference prediction, resource allocation, power optimization, and mobility prediction. Further research avenues include the convergence of licensed and unlicensed spectra, as well as leveraging machine learning techniques to enhance the efficiency of wireless networks by exploiting spectrum availability in both frequency bands.

3. **Future of ML in Network Layer:** In addition to the separation of network control from the data forwarding plane, there are crucial considerations regarding the application of ML. Specifically, extensive research has explored the integration of ML into routing protocols for mobile ad-hoc networks (MANETs).

Moreover, the security of these learning algorithms poses a significant challenge. In decentralized systems, the dissemination of misleading or incorrect information by sporadic devices can adversely impact the overall behavior of the network or routing protocols. Consequently, the utilization of ML in MANETs based on latency or time constraints and ensuring the security of the network present two compelling research challenges.

The research challenges extend beyond MANETs, encompassing nearly all types of networks, where routing protocols face issues such as network overhead due to information retrieval for learning, latency in making learning-based decisions, and ensuring the security of the learning environment. Consequently, these challenges persist as compelling areas for further investigation in research.

4. **Future of ML for Edge Computing:** Potential directions for future research encompass the advancement of methodologies leveraging ML as a pivotal element in decision-making processes for the allocation of application and service data and processing. This involves determining which components should be managed locally, which are at the edge, and which should be directed to data centers, known as ML for fog computing. These investigations should take into account factors such as performance, efficiency, and privacy considerations. Additionally, an intriguing avenue for future exploration lies in utilizing ML as a central component for orchestrating the management of edge infrastructure. This involves coordinating the allocation of nodes to serve users/end devices in specific logical or physical areas, or in certain application types. Another noteworthy perspective for future research revolves around the utilization of extensive datasets from diverse sources.

5.7 CONCLUSION

This chapter presents a comprehensive examination of the current state-of-the-art in the fields of ML disciplines, techniques, and tools as applied to communication networks. The discussion begins by elucidating the diverse applications of ML within wireless networks, spanning the physical, MAC, and network layers. Subsequently, this chapter explores ML applications in innovative technological paradigms such as SDN, NFV, and MEC. A specific emphasis is placed on delineating the role of ML in network security, underscoring its imperative for bolstering security operations in wireless networks. This chapter underscores the confluence of significant research endeavors that bridge ML disciplines with communication network technologies. Given the nascent deployment of ML in wireless networks, this chapter identifies numerous intriguing research questions warranting further investigation. Consequently, it outlines compelling future research directions spanning from the physical layer to the network layer and encompassing emerging networking technologies and services.

REFERENCES

1. M. Ibnkahla, Neural network predistortion technique for digital satellite communications, in *2000 IEEE International Conference on Acoustics, Speech, and Signal Processing*, vol. 6, June 2000, pp. 3506–3509.
2. F. Langlet, H. Abdulkader, D. Roviras, A. Maller, and F. Castanie, Adaptive predistortion for solid state power amplifier using multilayer perceptron, in *GLOBECOM'01. IEEE Global Telecommunications Conference*, vol. 1, Nov 2001, pp. 325–329.
3. J. Bruck, and M. Blaum, Neural networks, error-correcting codes, and polynomials over the binary n-cube, *IEEE Transactions on Information Theory*, vol. 35, no. 5, pp. 976–987, 1989.
4. G. Zeng, D. Hush, and N. Ahmed, An application of neural net in decoding error-correcting codes, in *IEEE International Symposium on Circuits and Systems*, vol. 2, May 1989, pp. 782–785.
5. W. R. Caid, and R. W. Means, Neural network error correcting decoders for block and convolutional codes, in *[Proceedings] GLOBECOM '90: IEEE Global Telecommunications Conference and Exhibition*, vol. 2, Dec 1990, pp. 1028–1031.
6. F. Liang, C. Shen, and F. Wu, An iterative BP-CNN architecture for channel decoding, *IEEE Journal of Selected Topics in Signal Processing*, vol. 12, no. 1, pp. 144–159, 2018.
7. W. Lyu, Z. Zhang, C. Jiao, K. Qin, and H. Zhang, Performance evaluation of channel decoding with deep neural networks, in

2018 IEEE International Conference on Communications (ICC), May 2018, pp. 1–6.

8. H. L. Southall, J. A. Simmers, and T. H. O'Donnell, Direction finding in phased arrays with a neural network beamformer, *IEEE Transactions on Antennas and Propagation*, vol. 43, no. 12, pp. 1369–1374, 1995.
9. R. Lovato and X. Gong, Phased antenna array beamforming using convolutional neural networks, in *2019 IEEE International Symposium on Antennas and Propagation and USNC-URSI Radio Science Meeting*, July 2019, pp. 1247–1248.
10. R. K. Ganti and M. Haenggi, Spatial and temporal correlation of the interference in ALOHA ad hoc networks, *IEEE Communications Letters*, vol. 13, no. 9, pp. 631–633, 2009.
11. M. Ozawa, T. Ohtsuki, W. Jiang, and Y. Takatori, Interference alignment for time-varying channel with channel and weight predictions based on auto regressive model, in *IEEE Global Communications Conference (GLOBECOM)*, 2015, pp. 1–6.
12. S. Ezpeleta, J. Claver, J. Pérez-Solano, and J. Martí, RF-based location using interpolation functions to reduce fingerprint mapping, *Sensors*, vol. 15, no. 10, pp. 27322–27340, 2015.
13. W. Zhang, K. Liu, W. Zhang, Y. Zhang, and J. Gu, Deep neural networks for wireless localization in indoor and outdoor environments, *Neurocomputing*, vol. 194, pp. 279–287, 2016.
14. Y. Tifani, B. Lee, and E. Jeong, A patient's indoor positioning algorithm using artificial neural network and SVM, *Journal of Theoretical & Applied Information Technology*, vol. 95, no. 16, 2017.
15. B. Berruet, O. Baala, A. Caminada, and V. Guillet, DelFin: A deep learning based CSI fingerprinting indoor localization in IoT context, in *International Conference on Indoor Positioning and Indoor Navigation (IPIN)*. IEEE, 2018, pp. 1–8.
16. Q. Liao, Dynamic uplink/downlink resource management in flexible duplex-enabled wireless networks, in *2017 IEEE International Conference on Communications Workshops (ICC Workshops)*, 2017, pp. 625–631.
17. H. Malik, M. M. Alam, H. Pervaiz, Y. L. Moullec, A. Al-Dulaimi, S. Parand, and L. Reggiani, Radio resource management in NB-IoT systems: Empowered by interference prediction and flexible duplexing, *IEEE Network*, vol. 34, no. 1, pp. 144–151, 2020.

18. Y. Li, H. Hu, Y. Wen, and J. Zhang, Learning-based power prediction for data centre operations via deep neural networks, in *Proceedings of the 5th International Workshop on Energy Efficient Data Centres, ser. E2DC '16*. New York, NY, USA: ACM, 2016, pp. 6:1–6:10.
19. H. Feng and Y. Shu, Study on network traffic prediction techniques, in *International Conference on Wireless Communications, Networking and Mobile Computing*, vol. 2, 2005, pp. 1041–1044.
20. Z. Zhao, W. Chen, X. Wu, P. C. Y. Chen, and J. Liu, LSTM network: A deep learning approach for short-term traffic forecast, *IET Intelligent Transport Systems*, vol. 11, no. 2, pp. 68–75, 2017.
21. R. J. Perlman, Network layer protocols with byzantine robustness, Ph.D. dissertation, Massachusetts Institute of Technology, 1988.
22. A. Feldmann, A. Greenberg, C. Lund, N. Reingold, and J. Rexford, NetScope: traffic engineering for IP networks, *IEEE Network*, vol. 14, no. 2, pp. 11–19, March 2000.
23. R. Li, Z. Zhao, X. Zhou, J. Palicot, and H. Zhang, The prediction analysis of cellular radio access network traffic: From entropy theory to networking practice, *IEEE Communications Magazine*, vol. 52, no. 6, pp. 234–240, June 2014.
24. W. Juan, S. Min, Z. Yan, W. Xijun, and L. Yuzhou, Traffic characteristics based dynamic radio resource management in heterogeneous wireless networks, *China Communications*, vol. 11, no. 1, pp. 1–11, Jan 2014.
25. F. Morales, M. Ruiz, L. Gifre, L. M. Contreras, V. Lopez, and L. Velasco, Virtual network topology adaptability based on data analytics for traffic prediction, *IEEE/OSA Journal of Optical Communications and Networking*, vol. 9, no. 1, pp. A35–A45, Jan 2017.
26. M. Suchara, D. Xu, R. Doverspike, D. Johnson, and J. Rexford, "Network Architecture for Joint Failure Recovery and Traffic Engineering," in Proceedings of the ACM SIGMETRICS Joint International Conference on Measurement and Modeling of Computer Systems, ser. SIGMETRICS '11. New York, NY, USA: ACM, 2011, pp. 97–108. [Online]. Available: http://doi.acm.org/10.1145/1993744.1993756
27. Z. M. Fadlullah, F. Tang, B. Mao, N. Kato, O. Akashi, T. Inoue, and K. Mizutani, State-of-the-art deep learning: Evolving machine

intelligence toward tomorrow's intelligent network traffic control systems. *IEEE Communications Surveys & Tutorials*, 19(4), pp. 2432–2455, 2017.
28. F. Tang, B. Mao, Z. M. Fadlullah, N. Kato, O. Akashi, T. Inoue, and K. Mizutani, On removing routing protocol from future wireless networks: A real-time deep learning approach for intelligent traffic control, *IEEE Wireless Communications*, vol. 25, no. 1, pp. 154–160, Feb. 2018.
29. K. Adachi, M. Li, P. H. Tan, Y. Zhou, and S. Sun, Q-learning based intelligent traffic steering in heterogeneous network, in *2016 IEEE 83rd Vehicular Technology Conference (VTC Spring)*, 2016, pp. 1–5.
30. A. Valadarsky, M. Schapira, D. Shahaf, and A. Tamar, A machine learning approach to routing, *arXiv preprint arXiv:1708.03074*, 2017.
31. N. Feamster, J. Borkenhagen, and J. Rexford, Guidelines for interdomain traffic engineering, *SIGCOMM Computer Communication Review*, vol. 33, no. 5, pp. 19–30, Oct. 2003. [Online]. Available: http://doi.acm.org/10.1145/963985.963988
32. M. Salani, C. Rottondi, and M. Tornatore, Routing and spectrum assignment integrating machine-learning-based QOT estimation in elastic optical networks, in *IEEE INFOCOM 2019 – IEEE Conference on Computer Communications*, April 2019, pp. 1738–1746.
33. F. Musumeci, C. Rottondi, A. Nag, I. Macaluso, D. Zibar, M. Ruffini, and M. Tornatore, An overview on application of machine learning techniques in optical networks. *IEEE Communications Surveys & Tutorials*, 21(2), pp. 1383–1408, 2018.
34. D. K. Sharma, S. K. Dhurandher, I. Woungang, R. K. Srivastava, A. Mohananey, and J. J. P. C. Rodrigues, A machine learning-based protocol for efficient routing in opportunistic networks, *IEEE Systems Journal*, vol. 12, no. 3, pp. 2207–2213, Sept 2018.
35. T. Huang, C. Lee, and L. Chen, PRoPHET+: An adaptive PRoPHETBased routing protocol for opportunistic network, in *2010 24th IEEEInternational Conference on Advanced Information Networking and Applications*, April 2010, pp. 112–119.
36. L. Zhi-yuan, W. Ru-chuan, and B. Jun-lei, A multipath routing algorithm based on traffic prediction in wireless mesh networks, in *2009 Fifth International Conference on Natural Computation*, vol. 6, Aug 2009, pp. 115–119.

37. N. Kato, Z. M. Fadlullah, B. Mao, F. Tang, O. Akashi, T. Inoue, and K. Mizutani, The deep learning vision for heterogeneous network traffic control: Proposal, challenges, and future perspective, *IEEE Wireless Communications*, vol. 24, no. 3, pp. 146–153, June 2017.
38. Q. Pham, F. Fang, V. N. Ha, M. J. Piran, M. Le, L. B. Le, W. Hwang, and Z. Ding, A survey of multi-access edge computing in 5G and beyond: Fundamentals, technology integration, and state-of-the-art, *IEEE Access*, vol. 8, pp. 116 974–117 017, 2020.
39. J. Park, S. Samarakoon, M. Bennis, and M. Debbah, Wireless network intelligence at the edge. *Proceedings of the IEEE*, 107(11), pp. 2204–2239, 2019.
40. R. G. Aryal and J. Altmann, Dynamic application deployment in federations of clouds and edge resources using a multiobjective optimization AI algorithm, in *2018 Third International Conference on Fog and Mobile Edge Computing (FMEC)*, April 2018, pp. 147–154.
41. S. B. Calo, M. Touna, D. C. Verma, and A. Cullen, Edge computing architecture for applying AI to IoT, in *2017 IEEE International Conference on Big Data (Big Data)*, Dec 2017, pp. 3012–3016.
42. S. Wang, M. Chen, X. Liu, C. Yin, S. Cui, and H. V. Poor, A machine learning approach for task and resource allocation in mobile edge computing based networks, *IEEE Internet of Things Journal*, vol. 8, pp. 1–1, 2020.
43. T. K. Rodrigues, K. Suto, H. Nishiyama, J. Liu, and N. Kato, Machine learning meets computation and communication control in evolving edge and cloud: Challenges and future perspective. *IEEE Communications Surveys & Tutorials*, 22(1), pp. 38–67, 2019.
44. H. Wu, Z. Zhang, C. Guan, K. Wolter, and M. Xu, Collaborate edge and cloud computing with distributed deep learning for smart city Internet of Things, *IEEE Internet of Things Journal*, vol. 7, no. 9, pp. 8099–8110, 2020.
45. M. Nduwayezu, Q. Pham, and W. Hwang, Online computation offloading in NOMA-based multi-access edge computing: A deep reinforcement learning approach, *IEEE Access*, vol. 8, pp. 99 098–99 109, 2020.
46. W. Jiang, G. Feng, S. Qin, and Y. Liu, Multi-agent reinforcement learning based cooperative content caching for mobile edge networks, *IEEE Access*, vol. 7, pp. 61 856–61 867, 2019.

47. Z. Cao, P. Zhou, R. Li, S. Huang, and D. Wu, Multiagent deep reinforcement learning for joint multi-channel access and task offloading of mobile-edge computing in Industry 4.0, *IEEE Internet of Things Journal*, vol. 7, no. 7, pp. 6201–6213, 2020.
48. Q. Cui, Z. Gong, W. Ni, Y. Hou, X. Chen, X. Tao, and P. Zhang, Stochastic online learning for mobile edge computing: Learning from changes, *IEEE Communications Magazine*, vol. 57, no. 3, pp. 63–69, 2019.
49. T. Subramanya, D. Harutyunyan, and R. Riggio, Machine learning-driven service function chain placement and scaling in MEC-enabled 5G networks, *Computer Networks*, vol. 166, p. 106980, 2020. [Online]. Available: http://www.sciencedirect.com/science/article/pii/S1389128619310254
50. M. Chen, Z. Yang, W. Saad, C. Yin, H. V. Poor, and S. Cui, A joint learning and communications framework for federated learning over wireless networks, *arXiv preprint arXiv:1909.07972*, 2019.
51. N. Shlezinger, M. Chen, Y. C. Eldar, H. V. Poor, and S. Cui, Federated learning with quantization constraints, in *ICASSP 2020 - 2020 IEEE International Conference on Acoustics, Speech and Signal Processing (ICASSP)*, 2020, pp. 8851–8855.
52. H. B. McMahan, E. Moore, D. Ramage, S. Hampson, and B. A. y Arcas, Communication-efficient learning of deep networks from decentralized data, in *AISTATS*, 2017.
53. M. Chen, H. V. Poor, W. Saad, and S. Cui, Convergence time optimization for federated learning over wireless networks, *arXiv preprint arXiv:2001.07845*, 2020.
54. M. Chen, H. V. Poor, W. Saad, and S. Cui, Convergence time minimization of federated learning over wireless networks, in *ICC 2020 - 2020 IEEE International Conference on Communications (ICC)*, 2020, pp. 1–6.
55. J. Konecný, H. B. McMahan, D. Ramage, and P. Richtárik, "Federated optimization: Distributed machine learning for on-device intelligence," *CoRR*, vol. abs/1610.02527, 2016.
56. W. Y. B. Lim, N. C. Luong, D. T. Hoang, Y. Jiao, Y. C. Liang, Q. Yang, D. Niyato, and C. Miao, Federated learning in mobile edge networks: A comprehensive survey. *IEEE Communications Surveys & Tutorials*, 22(3), pp. 2031–2063, 2020.

CHAPTER 6

Cuckoo Search Optimizer Based on Aquila Exploration Strategy

Megha Varshney, Pravesh Kumar, Pinkey Chauhan, and Geeta Arora

6.1 INTRODUCTION

The family of optimization algorithms known as metaheuristic algorithms, which draw their inspiration from natural events or processes, has attracted a lot of interest recently because of its capacity to effectively tackle challenging issues. Even in situations when conventional optimization approaches may be inefficient or unfeasible, these algorithms are made to search and explore enormous solution spaces in pursuit of optimal or nearly optimal solutions [1].

Because they may be used in a variety of problem areas, including but not limited to engineering, finance, healthcare, logistics, and scheduling, metaheuristic algorithms are renowned for their adaptability and versatility. Genetic algorithms, particle swarm optimization, ant colony optimization, simulated annealing, and cuckoo search are a few well-known examples of metaheuristic algorithms.

The capacity of metaheuristic algorithms to quickly and roughly solve difficult problems is one of the key factors contributing to their popularity.

DOI: 10.1201/9781003472506-6

In contrast to exact optimization approaches, which can demand time-consuming searches or difficult mathematical formulations, metaheuristic algorithms can produce near-optimal solutions rapidly and at a cheap computing cost. They are thus suited for real-world situations where it may be computationally impractical or time-consuming to determine the precise optimal solution.

The solution space can also be explored and used by metaheuristic algorithms in a non-deterministic, adaptive way. They have the ability to avoid local optima, which are poor solutions that trap conventional optimization techniques, and to modify their search strategies in response to the nature of the problem and the environment.

Additionally, because metaheuristic algorithms frequently offer many answers, decision-makers can select from a variety of possibilities that trade-off various objectives. This helps decision-makers in multi-objective optimization situations make more informed decisions by giving them flexibility and insight into the trade-offs between competing objectives.

Within the domain of metaheuristics, the cuckoo search (CS) algorithm stands out for its unique approach inspired by the brood parasitic behavior of certain cuckoo bird species [2,3]. Introduced by Yang and Deb [3], CS has gained significant traction due to its effectiveness in tackling continuous and even NP-hard (non-deterministic polynomial time) problems. The core principles of CS mimic the cuckoo's strategy of laying eggs in other birds' nests. In the context of optimization, each nest represents a potential solution, and the cuckoo eggs symbolize new solutions. The algorithm incorporates three idealized rules:

- **Lévy Flight for Exploration:** Cuckoos employ a Lévy flight foraging strategy, characterized by long jumps interspersed with short exploratory movements. This mechanism allows CS to explore the search space effectively, avoiding premature convergence on local optima [4].
- **Discovery and Replacement:** A fraction of host birds (existing solutions) may discover the foreign cuckoo eggs (inferior solutions). These discovered eggs are then abandoned or replaced with new ones, promoting the survival of better solutions.
- **Selection and Improvement:** The solutions (nests) with higher fitness values are carried over to the next generation, gradually leading to the identification of the optimal solution.

6.1.1 Evolution of CS and Its Limitations

Despite its success, CS has faced criticism for its susceptibility to premature convergence, particularly in complex problems with numerous local optima [5,6]. Researchers have addressed this limitation by proposing various enhancements to the original CS framework. These improved variants incorporate diverse strategies like hybrid approaches with other metaheuristics, parameter tuning techniques, and adaptive search mechanisms [7,8,9].

In 2021, Abualigah and Elaziz unveiled the revolutionary metaheuristic optimization technique known as the Aquila optimizer. It is intended to solve optimization problems effectively by emulating aquilas' foraging techniques. The exploration capabilities of the algorithm are improved by the exploration method used by the Aquila optimizer, allowing it to seek for optimal or nearly optimal solutions in challenging problem domains. The method employs a series of operators, such as the capturing, exploring, and escape operators, inspired by the hunting behavior of Aquilas, to repeatedly update the solution space and enhance the quality of solutions across iterations. The Aquila optimizer has drawn interest as a potential metaheuristic algorithm in the field of optimization research because of evidence of its promising performance in a variety of optimization tasks.

Here are some potential benefits of using the Aquila optimizer exploration strategy in the CS algorithm:

- **Enhanced Exploration:** The Aquila optimizer is known for its robust exploration capabilities. Its position update strategy, inspired by the hunting behavior of Aquila birds, allows for a wider search of the solution space. This can be particularly beneficial for CS, which can sometimes struggle with premature convergence, getting stuck in local optima. By incorporating Aquila's exploration methods, the combined Aquila-cuckoo search (ACS) algorithm could explore a more diverse range of solutions, increasing the chances of finding the global optimum.
- **Improved Search Diversity:** The Aquila optimizer utilizes various operators like "exploration" and "escape" that encourage movement away from potentially good solutions and toward unexplored areas.

This can help ACS maintain a healthy level of diversity in the population of solutions, preventing them from clustering around local optima too quickly. A diverse population allows for a more thorough exploration of the search space.

- **Reduced Premature Convergence:** As mentioned earlier, a major drawback of CS is its tendency to converge prematurely. By integrating Aquila's exploration strategy, ACS can potentially avoid getting trapped in local optima. The exploration operators in the Aquila strategy can help the algorithm escape these areas and continue searching for better solutions throughout the optimization process.
- **Potential for Faster Convergence:** While the primary focus of the Aquila strategy is exploration, it also incorporates elements of exploitation (focusing on promising areas). This combined approach might lead to a faster convergence on the optimal solution in ACS compared to the base CS algorithm. By efficiently exploring and then exploiting promising areas of the search space, ACS could find the optimal solution in fewer iterations.
- **Adaptability to Complex Problems:** The exploration strengths of the Aquila strategy can be particularly valuable for tackling complex optimization problems with many local optima. By leveraging Aquila's diverse search methods, ACS might be better equipped to navigate such challenging landscapes and locate the globally optimal solution.

It's important to note that these are potential benefits, and the actual effectiveness of the Aquila strategy in CS will depend on the specific problem being addressed and the implementation details of the ACS algorithm. However, the theoretical advantages suggest that the combination of these two algorithms has the potential to be a powerful tool for optimization tasks.

The study is set up like this: The operation of the Aquila optimizer algorithm and cuckoo search is explained in Section 6.2. The proposed ACS is described in Section 6.3. Parameter configurations, simulation outcomes, and comparisons with other algorithms are covered in Section 6.4. The conclusion is found in Section 6.5.

6.2 WORKING ON BASIC ALGORITHMS

This section explains the detailed functioning of the cuckoo search algorithm and the Aquila exploration strategy.

6.2.1 Cuckoo Search Algorithm

Levy's flight-based compulsory parasitism of some cuckoo species serves as the foundation for the CS algorithm.

This is how the CS process is explained:

- A cuckoo deposits an egg in a separate nest each time it does so.
- The best nest that yields an egg is passed down to the following generation.
- A limited quantity of host nests are accessible.

There is a likelihood $P_a(0,1)$ that host birds will find cuckoo eggs. The host bird in this scenario has two choices: it can either abandon the nest and build a new one or discard the eggs. In this instance, each egg can be considered a possible fix.

In the first stage of the CS process, when creating the ith solution in the $(k+1)$th generation, each solution is formed at random. The location of the nest or egg is updated by Eq. (6.1).

$$x_i(k+1) = x_i(k) + \alpha \otimes L(\lambda) \tag{6.1}$$

where $x_i(k+1)$ is the nest's ith position of the (k + 1) generation in the population, α is a real number, it marks the step size, $\otimes$ represents entry-wise multiplications, and $L(\lambda)$ is the random search vector produced by Levy's distribution.

6.2.2 Aquila Exploration Strategy

There exist two methods for the exploration of Aquila [10].

Exploration 6.1: Expanded Exploration (X_1)

This method involves the Aquila flying over the search region at a higher altitude, exploring a large area, and then performing a vertical dive to find the prey once the target location has been identified. The following is how this behavior can be stated mathematically:

$$X_1(k+1) = X_{\text{best}}(k)\left(1-\frac{k}{T}\right)+\left(X_M(k)-X_{\text{best}}(k)*\text{rand}\right) \quad (6.2)$$

$$X_M(k) = \frac{1}{N}\sum_{i=1}^{N} X_i(k), \quad \forall i = 1,2,\ldots,Dim \quad (6.3)$$

In this equation, $X_{\text{best}}(g)$ stands for the best position attained up until the current iteration, $X_M(g)$ stands for the average position of all Aquilas in the current iteration g, and $g_{\max}$ stands for the maximum number of iterations. N denotes the size of the population, while r is a chance number between 0 and 1.

Exploration 6.2: Narrowed Exploration (X_2)
Aquilas frequently use this hunting technique in which they swoop down in the intended region while using brief gliding, circle the prey, and then attack. The position update mathematical formula is written as:

$$X_2(k+1) = X_{\text{best}}(k)\times\text{Levy}(\lambda)+X_R(k)+(y-x)\times\text{rand} \quad (6.4)$$

where $X_R(g)$ stands for an Aquila location that is created at random, lambda stands for the dimension size, and rand is a random number between 0 and 1. The Levy flight, which has the following mathematical representation, is represented by the Levy(λ) function:

$$\text{Levy}(\lambda) = s\times\frac{u\times\sigma}{|v|^{\frac{1}{\beta}}} \quad (6.5)$$

$$\sigma = \left(\frac{\Gamma(1+\beta)\times\sin\left(\frac{\pi\beta}{2}\right)}{\Gamma\left(\frac{1+\beta}{2}\right)\times\beta\times 2^{\left(\frac{\beta-1}{2}\right)}}\right) \quad (6.6)$$

u and v are random numbers ranging from 0 to 1, and s and β are fixed constant values with 0.01 and 1.5, respectively. During the search, the variables y and x are used to indicate the spiral shape, and the following equations provide their calculation:

$$\begin{pmatrix} x = r\sin\theta \\ y = r\cos\theta \\ r = r_1 + U \times D_1 \\ \theta = -\omega \times D_1 + \theta_1 \\ \theta_1 = 3\pi/2 \end{pmatrix} \tag{6.7}$$

where r_1 represents the number of search cycles ranging from 1 to 20, D_1 comprises integer numbers from 1 to the dimension size (λ), and ω is set to 0.005.

6.3 PROPOSED ACS ALGORITHM

In accordance with the previously provided definition, we put forth our algorithm, known as ACS [11]. The two methods were modified as follows:

This program aims to identify the best solution for a given problem by iteratively updating a population of cuckoo eggs using the CS optimization algorithm and the Aquila exploration strategy. In order to determine whether cuckoo eggs are in an Aquila's catch range or escape range, the bird will adjust its position accordingly. Levy flying is another technique that the algorithm uses to produce fresh cuckoo eggs. The term "capture range" describes the area where an Aquila can detect and capture a cuckoo egg while it is in the exploring phase. If an Aquila is detected to be in range to grab a cuckoo egg, Eq. (6.2) is applied to update the cuckoo egg's position in relation to the direction and speed of the Aquila. The term "escape range," on the other hand, describes the area beyond which an Aquila is unlikely to accidentally trap a cuckoo egg. A cuckoo egg's position is updated using an escaping operator if it is discovered to be within Aquila's escape zone by using Eq. (6.4). The CS algorithm's Aquila exploration method takes into account both capture and escape range when considering how to update cuckoo eggs based on their proximity to an Aquila. The pseudocode of the proposed algorithm is given in **Algorithm 6.1**.

Core Principles:

- **Leveraging Cuckoo Search (CS):** ACS builds upon the foundation of the CS algorithm. It utilizes an initial population of solutions represented as "cuckoo eggs." These eggs are iteratively updated through the optimization process to find the solution with the best fitness value (optimal solution).
- **Enhancing Exploration with Aquila Strategy:** The key innovation lies in integrating the Aquila exploration strategy into CS. This strategy, inspired by the hunting behavior of Aquila birds, injects robust exploration mechanisms into the algorithm.

Exploration and Update Mechanisms:

- **Capture Range and Eq. (6.2):** The "capture range" defines the area where an Aquila bird can detect and potentially capture a cuckoo egg during its exploration phase. If a cuckoo egg falls within this capture range, Eq. (6.2) (not provided in the excerpt) is applied. This equation likely updates the cuckoo egg's position based on the Aquila's movement (direction and speed). This mechanism effectively guides the cuckoo egg toward potentially promising areas identified by the Aquila's exploration.
- **Escape Range and Eq. (6.4):** The "escape range" represents the zone beyond which an Aquila is unlikely to capture a cuckoo egg. If a cuckoo egg is detected within this escape range, Eq. (6.4) (not provided) comes into play. This equation likely involves an "escape operator" that updates the cuckoo egg's position away from the Aquila. This mechanism prevents good solutions (cuckoo eggs) from being prematurely discarded due to Aquila's presence.
- **Random Walk:** In scenarios where the cuckoo egg is neither within the capture nor escape range of an Aquila, the pseudocode suggests a random walk operator with probability pd. This mechanism ensures continued exploration of the search space even in the absence of direct Aquila influence.

Algorithm 6.1 Pseudo code of ACS

START
Input: Problem, population size, number of iterations up to maximum, p_a, and Aquila exploration method parameters.
Initialize: create the initial cuckoo population, When the maximum number of iterations has been reached, set the iteration counter to 0.
While (iteration counter < maximum number of iterations) **do**
  Examine the fitness of cuckoo eggs.
  For each cuckoo egg in the population, sort the eggs according to fitness do
    Use the Aquila exploring approach:
    If an Aquila is able to capture the cuckoo egg
    If yes, use Eq. (6.2) to update the cuckoo egg's position based on the direction and speed of the Aquila.
    If not, determine whether an Aquila's escape range includes the cuckoo egg.
    If yes, use Eq. (6.4)'s escape operator to update the cuckoo egg's position. If not, use the random walk operator with probability p_d.
  Produce a fresh cuckoo egg by using Levy flying.
    Analyze the new cuckoo egg's fitness
    Replace the cuckoo egg with the new cuckoo egg with probability p_a.
  END For
  Organize cuckoo eggs according to fitness
  Update the iteration counter
END while

Algorithm Procedure:

1. **Initialization:** The algorithm begins by receiving problem parameters, population size, maximum iterations, a parameter pa (likely controlling the replacement probability), and parameters specific to the Aquila exploration method. An initial population of cuckoo eggs (potential solutions) is then created.
2. **Iteration Loop:** The algorithm enters a loop that continues for a predefined number of iterations. Within each iteration:
   - o **Fitness Evaluation:** The fitness of each cuckoo egg (solution) in the population is evaluated.
   - o **Aquila Exploration and Update:** For each cuckoo egg:
     - The Aquila exploration strategy is applied. The egg's position is potentially updated based on its proximity to an Aquila (using capture range, Eq. (6.2), escape range, and Eq. (6.4)).

- o **Lévy Flight for New Eggs:** A new cuckoo egg (potential solution) is generated using the Lévy flight technique, a mechanism from the original CS algorithm known for its ability to promote exploration over large distances.
- o **Fitness Evaluation and Replacement:** The fitness of the newly generated egg is evaluated. With a certain probability (p_a), the new egg replaces the existing cuckoo egg if it has a better fitness value.

3. **Selection and Update:** After processing all eggs in the population, they are likely sorted based on their fitness values. The iteration counter is incremented.
4. **Termination:** The loop continues until the maximum number of iterations is reached.

Overall, the ACS algorithm merges the strengths of CS (exploration and exploitation) with the powerful exploration capabilities of the Aquila optimizer. By strategically updating cuckoo egg positions based on Aquila's exploration strategy, ACS aims to achieve a more thorough exploration of the search space and prevent premature convergence on local optima.

6.4 SIMULATION RESULTS AND DISCUSSION

We use a few well-known standard test beds to evaluate the proposed ACS algorithm. Table 6.1 displays the search limits and characterizations of these test issues. The algorithms mentioned above are all minimizations. From F1 to F4, functions are unimodal and scalable; from F5 to F9, functions are multimodal, and fixed-dimensional multimodal F10.

- **System Settings**: The algorithm is implemented using MATLAB. Additionally, the system operates on an Intel Core i5 processor clocked at 3.20 GHz and runs on a 64-bit Windows 7 operating system.
- **Parameter Settings:** The parameter settings of the algorithms used in this study are given in Table 6.1.
- **Benchmark Functions:** For evaluating ACS with CS and Aquila as well as comparison with other algorithms, the following benchmark functions were picked from the literature [12–16] and are given in Table 6.2.

TABLE 6.1 Parameter Settings of the Algorithms

Algorithm	Parameter	Value
CS	P_a	0.25
AO	α, β	0.1
PSO	I. W. Velocity	Decrement from 0.9 to 0.1 10% of dimension range
GWO	a	Decrement from 2 to 0
WOA	w_1 w_2 v	$[2,0]$ $[-1,-2]$ 1
SM A	vb and vc	Reduced from 2 to 0

TABLE 6.2 Benchmark Functions, Property, and Their Range

Function	Name	Property	Range
F1	Sphere function	Unimodal	[−100,100]
F2	Schwefel's 1.20 function	Unimodal	[−100,100]
F3	Schwefel's 2.21 function	Unimodal	[−30,30]
F4	Rosenbrock function	Unimodal	[−100,100]
F5	Schwefel's 2.26 function	Multimodal	[−500,500]
F6	Rastrigin's function	Multimodal	[−5.12,5.12]
F7	Ackley's function	Multimodal	[−32,32]
F8	Griewank function	Multimodal	[−600,600]
F9	Pendized function	Multimodal	[−50,50]
F10	Kowalik's function	Fixed-dimensional multimodal	[−5,5]

6.4.1 ACS's Intensification Capability

The results from Table 6.3 show the comparison of results between the basic algorithms AO and CS with ACS. And the obtained results show that ACS performs better than the original CS. Therefore, the addition of Aquila's exploration capability with cuckoo search gives excellent results. To evaluate the ability of ACS to utilize global solutions, we used a set of unimodal functions (F1–F4), which possess only one global optimum. The results obtained from assessing the unimodal benchmarks for ACS and other existing competitors are presented in Table 6.4 (F1–F4), including the worst and best solutions found so far, as well as the average and standard deviation values from several runs. Our findings support the superiority of ACS over its competitors, as it outperforms them in achieving minimum

TABLE 6.3 Results of the Algorithms AO, CS, and Proposed Algorithm ACS on Classical Benchmark Functions; the Dimension Is Fixed at 10

Function No.	Performance Measure	Aquila Optimizer (AO)	Cuckoo Search (CS)	ACS
F1	Worst	0.0×10^{0}	3.2×10^{-9}	0.0×10^{0}
	Average	0.0×10^{0}	9.7×10^{-10}	0.0×10^{0}
	Best	0.0×10^{0}	1.7×10^{-10}	0.0×10^{0}
	STD	0.0×10^{0}	7.8×10^{-10}	0.0×10^{0}
F2	Worst	0.0×10^{0}	5.9×10^{01}	0.0×10^{0}
	Average	0.0×10^{0}	2.8×10^{01}	0.0×10^{0}
	Best	0.0×10^{0}	6.2×10^{0}	0.0×10^{0}
	STD	0.0×10^{0}	1.8×10^{01}	0.0×10^{0}
F3	Worst	3.4×10^{-21}	2.8×10^{01}	2.4×10^{-08}
	Average	9.9×10^{-21}	1.0×10^{10}	2.1×10^{-09}
	Best	2.3×10^{-22}	8.8×10^{-02}	1.5×10^{-11}
	STD	0.0×10^{00}	6.5×10^{-01}	4.9×10^{-09}
F4	Worst	7.2×10^{-03}	$2.5\times10^{+01}$	2.1×10^{-01}
	Average	1.8×10^{-03}	$2.1\times10^{+01}$	1.7×10^{-03}
	Best	5.8×10^{-05}	$1.8\times10^{+01}$	1.7×10^{-01}
	STD	2.5×10^{-03}	$1.5\times10^{+00}$	1.4×10^{-01}
F5	Worst	1.5×10^{-03}	2.7×10^{-02}	1.4×10^{-04}
	Average	8.2×10^{-04}	1.3×10^{-02}	8.1×10^{-05}
	Best	3.6×10^{-05}	7.6×10^{-03}	2.5×10^{-06}
	STD	4.8×10^{-04}	5.2×10^{-03}	4.6×10^{-05}
F6	Worst	$^{-}1.4\times10^{+03}$	$^{-}6.4\times10^{+03}$	$^{-}1.3\times10^{+03}$
	Average	$^{-}2.2\times10^{+03}$	$^{-}7.1\times10^{+03}$	$^{-}1.1\times10^{+03}$
	Best	$^{-}2.7\times10^{+03}$	$^{-}7.7\times10^{+03}$	$^{-}1.2\times10^{+03}$
	STD	$4.7\times10^{+02}$	$3.3\times10^{+02}$	$4.4\times10^{+02}$
F7	Worst	0.0×10^{0}	$5.2\times10^{+01}$	0.0×10^{0}
	Average	0.0×10^{0}	$4.0\times10^{+01}$	0.0×10^{0}
	Best	0.0×10^{0}	$2.5\times10^{+01}$	0.0×10^{0}
	STD	0.0×10^{0}	$6.9\times10^{+00}$	0.0×10^{0}
F8	Worst	8.7×10^{-16}	3.3×10^{-01}	5.5×10^{-16}
	Average	8.5×10^{-16}	3.0×10^{-02}	7.3×10^{-11}
	Best	8.9×10^{-16}	6.0×10^{-05}	1.4×10^{-12}
	STD	0.0×10^{0}	6.9×10^{-02}	1.2×10^{-10}
F9	Worst	0.0×10^{0}	7.5×10^{-03}	0.0×10^{0}
	Average	0.0×10^{0}	2.4×10^{-04}	0.0×10^{0}
	Best	0.0×10^{0}	9.4×10^{-10}	0.0×10^{0}
	STD	0.0×10^{0}	1.2×10^{-03}	0.0×10^{0}
F10	Worst	8.4×10^{-04}	$2.0\times10^{+01}$	6.9×10^{-04}
	Average	5.5×10^{-04}	$1.7\times10^{+01}$	3.4×10^{-04}
	Best	4.0×10^{-04}	1.2×10^{01}	1.1×10^{-05}
	STD	1.9×10^{-04}	1.9×10^{00}	1.4×10^{-04}
	Mean	2.0×10^{00}	2.1×10^{00}	1.2×10^{00}
	Ranking	2	3	1

TABLE 6.4 Results of Other Optimization Algorithms for Solving the Classical Benchmark Functions, at Dimension 10

Function No.	Performance Measure	Different Algorithms				
		ACS	PSO	GWO	WOA	SMA
F1	Worst	0.0×10^{0}	6.2×10^{-9}	1.4×10^{-5}	2.9×10^{-7}	1.6×10^{-2}
	Average	0.0×10^{0}	1.6×10^{-9}	7.0×10^{-5}	1.0×10^{-7}	2.3×10^{-2}
	Best	0.0×10^{0}	2.8×10^{-2}	3.3×10^{-5}	1.2×10^{-7}	$0.0\times10^{+0}$
	STD	0.0×10^{0}	2.5×10^{-2}	5.5×10^{-5}	1.4×10^{-7}	$0.0\times10^{+0}$
F2	Worst	0.0×10^{0}	8.9×10^{-1}	5.9×10^{-2}	9.1×10^{-2}	5.9×10^{-2}
	Average	0.0×10^{0}	2.4×10^{-5}	1.4×10^{-2}	4.1×10^{-2}	8.4×10^{-2}
	Best	0.0×10^{0}	3.1×10^{-7}	5.8×10^{-2}	1.8×10^{-2}	0.0×10^{0}
	STD	0.0×10^{0}	3.7×10^{-5}	2.5×10^{-2}	3.7×10^{-2}	0.0×10^{0}
F3	Worst	2.4×10^{-8}	9.0×10^{-4}	5.9×10^{-5}	1.0×10^{-1}	4.1×10^{-1}
	Average	2.1×10^{-9}	2.4×10^{-4}	2.1×10^{-5}	3.1×10^{-0}	6.2×10^{-1}
	Best	1.5×10^{-1}	2.0×10^{-5}	2.3×10^{-6}	2.1×10^{-3}	4.1×10^{-2}
	STD	4.9×10^{-9}	3.7×10^{-4}	2.5×10^{-5}	4.0×10^{-0}	1.5×10^{-2}
F4	Worst	2.0×10^{-1}	7.6×10^{-0}	8.0×10^{-0}	8.9×10^{-0}	8.2×10^{-0}
	Average	1.8×10^{-3}	4.9×10^{-0}	6.9×10^{-0}	7.6×10^{-0}	6.6×10^{-0}
	Best	1.7×10^{-1}	1.0×10^{-1}	6.2×10^{-0}	6.8×10^{-0}	1.4×10^{-1}
	STD	1.3×10^{-1}	2.8×10^{-1}	7.6×10^{-1}	8.2×10^{-0}	2.9×10^{-2}
F5	Worst	1.4×10^{-4}	1.6×10^{-2}	1.3×10^{-3}	8.9×10^{-3}	9.9×10^{-4}
	Average	8.1×10^{-5}	9.4×10^{-3}	6.4×10^{-4}	5.2×10^{-3}	4.9×10^{-4}
	Best	2.5×10^{-6}	3.9×10^{-3}	3.9×10^{-4}	1.9×10^{-3}	6.3×10^{-5}
	STD	4.6×10^{-5}	4.3×10^{-3}	3.8×10^{-4}	3.3×10^{-3}	3.9×10^{-4}
F6	Worst	-1.2×10^{-3}	-1.5×10^{-3}	-2.8×10^{-3}	-2.3×10^{-3}	-4.1×10^{-3}
	Average	-1.2×10^{-3}	-2.1×10^{-3}	-2.7×10^{-3}	-3.1×10^{-3}	-4.1×10^{-3}
	Best	-1.2×10^{-3}	-2.5×10^{-3}	-3.4×10^{-3}	-3.8×10^{-3}	-4.1×10^{-3}
	STD	4.4×10^{-2}	4.4×10^{-3}	3.4×10^{-2}	4.5×10^{-2}	1.8×10^{-1}
F7	Worst	0.0×10^{0}	1.3×10	1.5×10^{-1}	1.4×10^{-4}	0.0×10^{0}
	Average	0.0×10^{0}	5.6×10	4.3×10^{-0}	2.8×10^{-5}	0.0×10^{0}
	Best	0.0×10^{0}	1.9×10	0.0×10^{-0}	0.0×10^{-0}	0.0×10^{0}
	STD	0.0×10^{0}	4.3×10	6.6×10^{-0}	6.3×10^{-5}	0.0×10^{0}
F8	Worst	5.5×10^{-6}	1.1×10^{-9}	1.5×10^{-4}	7.9×10^{-5}	8.8×10^{-6}
	Average	7.3×10^{-1}	7.0×10^{-0}	1.1×10^{-4}	4.4×10^{-5}	8.8×10^{-6}
	Best	1.4×10^{-2}	3.2×10^{-0}	7.9×10^{-5}	8.8×10^{-6}	8.8×10^{-6}
	STD	1.2×10^{-0}	3.7×10^{-0}	3.5×10^{-5}	3.5×10^{-5}	0.0×10^{-0}
F9	Worst	0.0×10^{0}	5.3×10^{-1}	5.9×10^{-2}	2.4×10^{-1}	0.0×10^{0}
	Average	0.0×10^{0}	2.4×10^{-1}	1.3×10^{-2}	9.7×10^{-2}	0.0×10^{0}
	Best	0.0×10^{0}	1×10^{-1}	0.0×10^{-0}	0.0×10^{-0}	0.0×10^{0}
	STD	0.0×10^{0}	1.7×10^{-1}	2.5×10^{-2}	1.3×10^{-1}	0.0×10^{0}
F10	Worst	6.9×10^{-4}	4.7×10^{-1}	1.0×10^{-2}	9.4×10^{-1}	1.0×10^{-2}
	Average	3.4×10^{-4}	3.9×10^{-1}	2.1×10^{-3}	5.6×10^{-1}	4.3×10^{-3}
	Best	1.1×10^{-5}	3.3×10^{-1}	5.7×10^{-0}	3.5×10^{-1}	2.2×10^{-7}
	STD	1.4×10^{-4}	5.0×10^{-2}	4.9×10^{-3}	2.3×10^{-1}	6.0×10^{-3}
	Mean	1.2×10^{-0}	5.9×10^{-0}	5.0×10^{-0}	6.2×10^{-0}	4.7×10^{-0}
	Ranking	1	4	3	5	2

values for the best and worst solutions found so far, as well as average values with high reliability and consistency (minimum STD) for six of the set's functions. However, ACS provides the least optimal number for F7. This insight reveals ACS's remarkable skill in enhancing the best solutions. The proposed ACS algorithm employs two exploitation strategies, namely expanded exploitation and narrowed exploitation, which primarily focus on the local search area. These strategies operate according to their fundamental procedures and facilitate the optimization process by explicitly and broadly targeting the neighborhood in the search.

6.4.2 ACS's Diversification Capability

Analyzing the numerous local optima of multimodal benchmarks allows one to assess the ability of metaheuristics (MHs) to explore problem spaces. In our study, we assessed the performance of ACS using both multidimensional (F5–F9) and fixed-dimensional (F10) multimodal test functions. The study findings for the multidimensional (F5–F9) and fixed-dimensional (F10) functions, as well as the worst, average, best, and standard deviation values of the functions analyzed, were provided in Table 6.4. Our investigation revealed that ACS outperformed its rivals on the majority of the high-dimensional multimodal benchmarks. In fact, for five of the six functions in this collection, ACS supplied the worst, average, best, and STD values. As seen in Table 6.4, the SMA was the sole substantial rival for ACS. Our findings show that generally, ACS outperformed its rivals. Due to its two exploration tactics, which concentrate on the worldwide search area, ACS has performed exceptionally well. These tactics enable ACS to explore the search space accurately and efficiently, outperforming more recently demonstrated algorithms. The process of optimization is managed in two distinct ways when searching for a large search region: widely and narrowly.

6.4.3 Analysis of the Qualitative Convergence of ACS

To validate the effectiveness of the ACS model, we utilized the convergence curves of standard functions such as F1, F2, and F10, as illustrated in Figure 6.1.

The second column of Figure 6.1 presents the convergence curve while the first column displays the function's 2D shape indicating qualitative measurements. The convergence curve depicted in the last column of Figure 6.1 shows various patterns associated with the evaluated functions. For instance, unimodal functions that meet arcs are aspherical and require

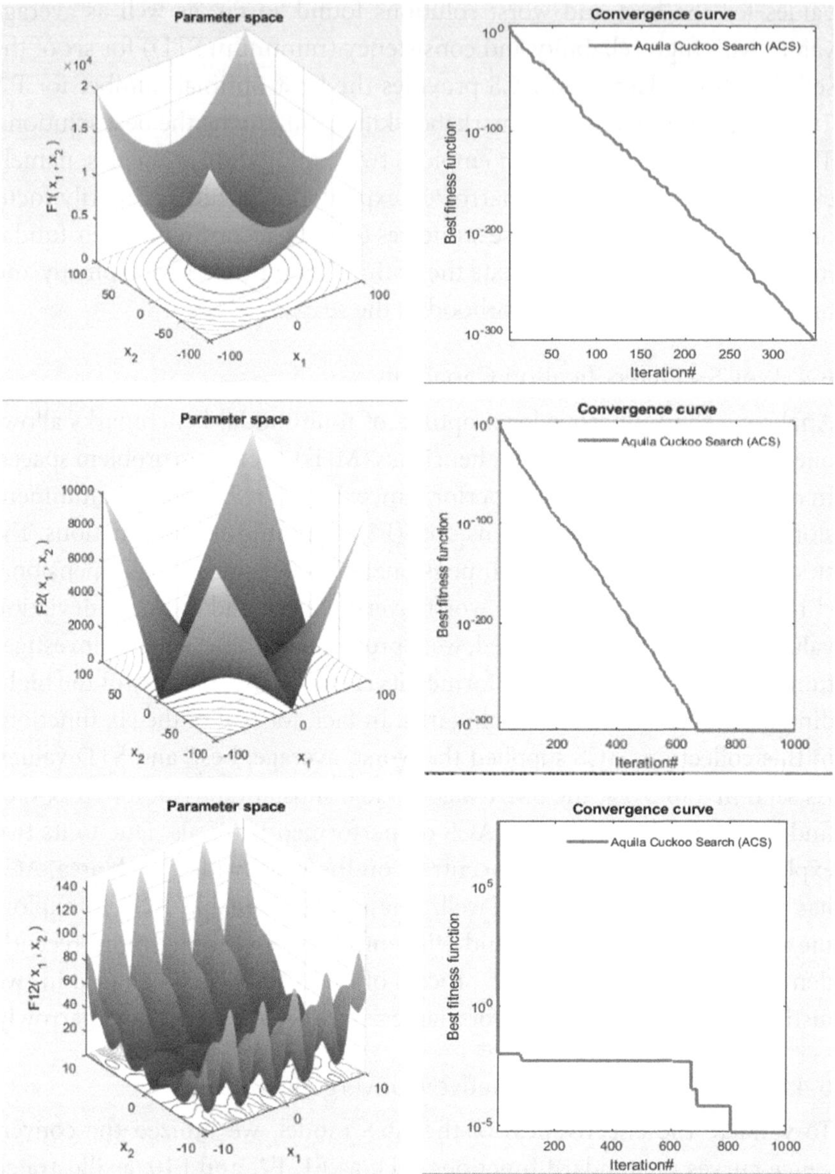

FIGURE 6.1 Convergence curve and parameter space of the functions F1, F2, and F12.

minor adjustments for better results. On the other hand, multimodal functions are more intricate and their convergence trajectory exhibits a stepwise pattern.

6.5 CONCLUSION

The study proposes a new hybrid optimization algorithm, ACS, which tackles a major limitation of CS - getting stuck on local solutions. ACS incorporates the exploration strengths of the Aquila optimizer into the CS framework. This allows ACS to effectively explore the search space and avoid premature convergence. This chapter showcases ACS's superiority. It will have the potential application in various domains like big data, text mining, image processing, and even neural network training. ACS will emerge as a significant advancement in optimization, offering a robust tool for real-world problems across diverse disciplines.

REFERENCES

1. Yang, X., Gandomi, A.H., Talatahari, S., Alavi, A.H., *Optimization and Metaheuristic Algorithm in Engineering*. Elsevier. ISBN 978-0-12-398296-4, 2013.
2. Pentapalli, V., Varma, R, Cuckoo search optimization and its applications: A review. *International Journal of Advanced Research in Computer and Communication Engineering*, 5(11), 556–562, November, 2016.
3. Yang, X.S., Deb, S., Cuckoo search via Lévy flights. *Proceedings of the World Congress on Nature and Biologically Inspired Computing-NaBIC*, Coimbatore, India, December, pp. 210–214, 2009.
4. Shehab, M., Khader, A.T., Al-Betar, M.A., A survey on applications and variants of the cuckoo search algorithm. *Applied Soft Computing*, 61, 1041–1059, September 2017.
5. Kamoona, M., Patra, J., Stojcevski, A., An enhanced cuckoo search algorithm for solving optimization problems. *2018 IEEE Congress on Evolutionary Computation (CEC)*, Rio de Janeiro, Brazil, July, 2018.
6. Al-Abaji, M.A., A literature review of Cuckoo search algorithm. *Journal of Education and Practice*, 11(8), 1–8, 2020.
7. Fister Jr, I., Fister, D., Fister, I., Cuckoo search: A brief literature review. *Part of Studies in Computational Intelligence Book Series*, 11, 49–62, 2020.

8. Yang, X.S., Deb, S., Cuckoo search recent advances and applications. *Neural Computing and Applications*, 24, March, 169–174, 2013.
9. Abualigah, L., Diabat, A., Mirjalili, S., Abd Elaziz, M., Gandomi, A.H., The arithmetic optimization algorithm. *Computer Methods in Applied Mechanics and Engineering*, 376, 113609, 2021.
10. Abualigah, L., Yousri, D., Abd Elaziz, M., Ewees, A.A., Al-qaness, M.A.A., Gandomi, A.H., Aquila optimizer: A novel meta-heuristic optimization algorithm. *Computers & Industrial Engineering*, 157, 107250, 2021.
11. Varshney, M., Kumar, P., Ali, M., Gulzar, Y. Using the Grey Wolf Aquila synergistic algorithm for design problems in structural engineering. *Biomimetics*, 9, 54, 2024. https://doi.org/10.3390/biomimetics9010054
12. Gupta, S., Deep, K., Heidari H.M.J.E.S., Wang, W.A., Opposition-based learning Harris hawks optimization with advanced transition rules principles and analysis. *Expert Systems with Applications*, 158, 113510, 2020.
13. Hakli, H., Uğuz, H., A novel particle swarm optimization algorithm with Levy flight. *Applied Soft Computing*, 23, 333–345, 2014.
14. Hatta, N.M., Zain, A.M., Sallehuddin, R., Shayfull, Z., Yusoff, Y. Recent studies on optimization method of Grey Wolf optimiser (GWO): A review (2014–2017). *Artificial Intelligence Review*, 52, 2651–2683, 2019.
15. Mirjalili, S., Mirjalili, S.M., Lewis, A., Grey Wolf optimizer. *Advances in Engineering Software*, 69, 46–61, 2014.
16. Precup, R.E., David, R.C., Roman, R.C., Petriu, E. M., Szedlak-Stinean, A.I., Slime mould algorithm-based tuning of cost-effective fuzzy controllers for the servo system. *International Journal of Computational Intelligence Systems*, 14, 1042–1052, 2019.

CHAPTER 7

Semantic Segmentation

Techniques, Challenges, and Applications in Intelligent Transportation Systems

Nouman Latif, Muhammad Saadi,
and Demostenes Zegarra Rodriguez

7.1 INTRODUCTION

Artificial intelligence is helping human beings in almost every field. For example, if we are using a mobile phone for taking pictures, artificial intelligence (AI) is used for refining it. You can see the AI working when using social media. You search about some topic and advertisements start appearing on your social media account about it. AI helps in detecting and preventing diseases. AI is playing a role in protecting the environment. It helps bankers, businessmen, and almost everyone in analysis to make different decisions. AI is starting to become an integral part of cars, aircraft, gadgets, medical devices, machines, etc. There are numerous applications of AI and it is progressing day by day. The field of communication is one of the many, which is being revolutionized by the use of AI. Communication is used in many applications, and the whole world is connected at different levels through communication systems. The Internet was revolutionary. It connects people through different devices. But now, with the progress in AI, not only humans are communicating but also machines are also communicating with each other. The communication between machines is called the Internet of things (IoT). This technology allows different sensors and

DOI: 10.1201/9781003472506-7

machines to share data. A variant of IoT is the Internet of vehicles (IoV). In IoV, different vehicles communicate with each other on roads and share information with each other which can be crucial for safety. This field has come into existence because of the progressing technology of autonomous vehicles. An autonomous driving system (ADS) requires information from different types of sensors and from other vehicles to make a decision. One of the most important and computationally extensive data is image which has a vital role in environment perception. Images are to be transmitted from sensor to vehicle controller in intra-vehicle communication, from one vehicle to other vehicles in vehicle-to-vehicle (V2V) communications, and from vehicle-to-infrastructure in vehicle-to-infrastructure (V2I) communication. Apart from intelligent transportation systems, transmitting images in any system is challenging. AI is also used to increase the efficiency of this task whether it is wired communication or wireless communication. The techniques used for this application utilize different deep learning approaches that deduct semantic information from the image, discard redundant information, and transmit only the semantic information through a communication channel. This chapter will discuss semantic segmentation, its types, techniques, and its implementation in communication systems for efficient and low-latency transmission of images.

7.2 IMAGE SEGMENTATION

Any system of AI needs information using which it can make decisions and perform any task such as predicting the price of a house, generating images, writing stories, determining whether a patient has a disease or not, guiding an autonomous vehicle, and predicting a cyber-attack. For the tasks in which information is required for an AI system to see the world like humans do, image segmentation is used. It is a fundamental technique in machine vision that is used to extract information from images or videos. The extracted information should be useful and it is possible if we divide an image into some segments or parts. In image segmentation, this division is performed on the pixels that belong to a certain group based on similarity between the pixels. For example, consider the image in Figure 7.1 which contains different objects or classes:

If we want an AI system to define this image, then it can do it efficiently if the information of the image is divided into parts such as grass, ropes, human, sheep, and a dog. Image segmentation will make groups of pixels

FIGURE 7.1 Original image with different objects such as person, sheep, dog, and background segments [1].

FIGURE 7.2 Image is segmented into different classes. Each class is highlighted with pixels of different colors [1].

in such a way that pixels belonging to a human will make one segment, pixels for a dog will make another segment, there will be a segment of sheep, and then for the grass. Using this useful coherent information about the segments in an image, now an AI system or a machine vision system can interpret this image. The image segmentation divides the image into meaningful parts by measuring the similarity of the pixels in terms of certain characteristics such as textures, intensity, or color. Figure 7.2 shows a segmented image in which different objects shown in the image of Figure 7.1 are labeled using different colors such as background pixels are labeled as green, pixels related to humans are sky blue, sheep are labeled with dark blue color, and the dog is represented using pixels of red color.

7.2.1 Types of Image Segmentation

Grouping the pixels in an image based on similarities between them is segmentation and depending on the methodology and scope of segmentation provides us with different types of segmentation. The following are the types of image segmentation:

7.2.1.1 Thresholding-Based Segmentation

An image consists of pixels, and each pixel has a value between 0 and 255. If it is a grayscale image, then there will be only one channel, and if it is a color image, then there will be three channels. In thresholding, if there is a substantial difference between the pixel values of one object and the background, a threshold value is decided. Pixels with values lower than the threshold will be grouped into one segment and pixels with values greater than the threshold will be assigned another segment. This technique is also called binarization of an image as there will be only two segments in the output. For example, all background pixels are converted into black pixels, and all pixels related to the region of interest are converted into black color pixels. This can be observed in Figure 7.3. In this figure, the region of interest is turned white and the background is converted as black pixels.

FIGURE 7.3 (a) Original image with different object classes (b) image after threshold based segmentation in which region of interest is represented by black pixels and background is represented by white pixels [2].

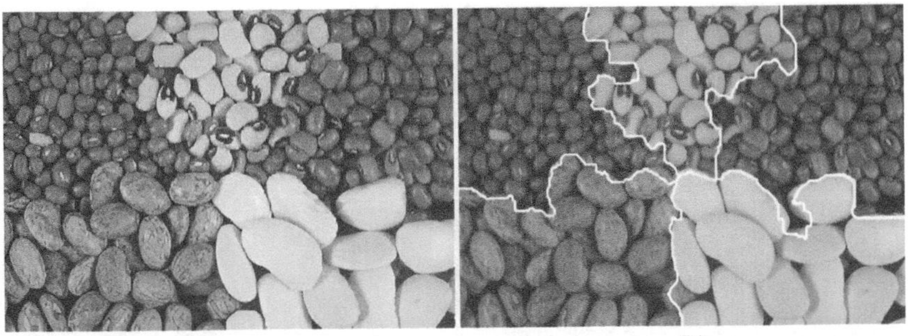

FIGURE 7.4 (a) Different types of beans are shown and the task is to separate different types; (b) using edge-based segmentation, different types of beans are separated with clear edges between them [3].

7.2.1.2 Edge-Based Segmentation

To differentiate different objects in an image, edge detection can be used to identify the boundaries. It works on the principle that when we pass through an edge in an image, an abrupt change is observed. This change can tell us if there is a boundary or edge at that instance. For example, if we want to differentiate a car from its background, then we can use edge detection. Consider Figure 7.4 to understand this concept. In Figure 7.4a, different types of beans are shown and we can observe boundaries between these types. Using edge-based detection, these boundaries are clearly differentiating types of beans in an image as shown in Figure 7.4b.

7.2.1.3 Cluster-Based Segmentation

In machine learning, clustering is a type of unsupervised learning in which the data has no labels, and based on the similarity between different data points, the clusters are made. For segmentation, this approach can be used for differentiating different objects in an image. This method is useful because the pixels in an image are also unlabeled. Different techniques are available for cluster-based segmentation. One of these methods is K-means clustering in which K-number clusters are made based on different values of centers equal to the number of clusters. For example, in Figure 7.5, the value for K is 2 which means the image is segmented into two clusters. One for the background and one for the class animals.

7.2.1.4 Graph-Based Image Segmentation

Segmentation is all about coherence between the pixels based on which different parts in an image are grouped into segments. One of the emerging

(a) (b)

FIGURE 7.5 Image segmentation performed using clustering with number of clusters 2. (a) Original image with cows and background; (b) segmentation of cows is shown with dark grey color pixels, and background is shown in light grey background pixels [4].

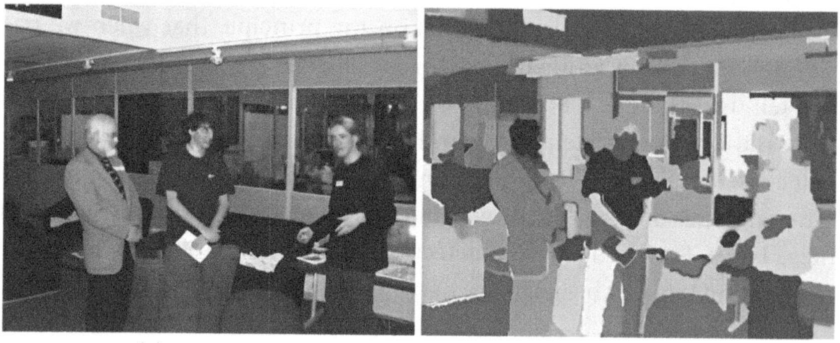

(a) (b)

FIGURE 7.6 Image segmentation using graphs (a) original image with multiple objects (b) segmented image using different pixel colors for different objects [5].

and fast-growing techniques in machine learning is graph neural networks (GNU). This method is used for various types of data but we can also use it for image segmentation. For this purpose, an image is converted into a graphical representation with pixels as nodes and connectivity showing coherence between the pixels by using edges. Figure 7.6 shows the segmentation of an image using graphs. The graph of the original image is divided into subgraphs, and based on the similarity between the pixels, different clusters are made as shown in Figure 7.6b.

7.2.1.5 Semantic Segmentation

The difference between the different types discussed previously and semantic segmentation is that labels are given to different objects in an image based

FIGURE 7.7 Semantic segmentation using fully convolutional network (FCN). (a) Original image containing different objects such as bikes, humans, road; (b) semantically segmented image in which different objects are represented using different colors [6].

on the similarity between the pixels. Moreover, the methodology used for this purpose is supervised in nature, such as a very famous technique called fully convolutional networks (FCN) in which labeled images are used to train a network to get the semantic information. Consider an example in Figure 7.7. In the original image, we can observe that there are riders on bikes and then there is a road and an audience. The semantic segmentation is shown in Figure 7.7b; it can be observed that pixels for bikes, humans, and backgrounds are grouped as different segments.

7.2.1.6 Instance-Based Segmentation

This type of segmentation has taken semantic segmentation one step ahead as it even differentiates or labels the pixels of the same class. To understand this concept, first consider Figure 7.7 in which humans are grouped in one cluster, background in one, and audience in another cluster. Instance segmentation will not only group the pixels based on similarity but also it can group clusters within clusters. It will also label the humans such as Mike riding a bike on the left and Jhon riding the bike on the right. Figure 7.8 shows an example of instance segmentation.

In Figure 7.8a, there are three humans and then there is a background with different objects. If we use instance segmentation to differentiate between the humans and the background, it will also group the humans in different clusters as shown by different colors in Figure 7.8b. If this was semantic segmentation, all humans would have been colored the same. So, instance segmentation gives more granular semantic information.

(a) (b)

FIGURE 7.8 Instance segmentation. (a) Original image with different objects; (b) Different humans are labeled differently [7].

(a) (b)

FIGURE 7.9 Image segmentation using deep learning. (a) Original image with different objects; (b) segmented image [8].

7.2.1.7 Deep Learning-Based Segmentation

In all of the previous types, we have discussed that different objects in an image are segmented based on their similarity. This similarity is between the texture or color or edges which represent the features of the objects. Usually, these features are first extracted using hand-crafted techniques, but in deep learning, an end-to-end pipeline works to automatically extract these features and then perform the segmentation. The deep learning approach is convolutional neural network (CNN) based. Figure 7.9 shows how the quality of segmentation is improved by adapting the deep learning approach.

7.3 INTELLIGENT TRANSPORTATION SYSTEM (ITS)

With an increasing population, the need for transport is also increasing. With this expansion in ground transportation, there is a need for enhanced safety, security, efficiency, and reliability. To handle different aspects of this system, ITS utilizes different technologies, including information, control, and communications. ITS is a vast field that covers various aspects of transport systems such as:

- Traffic management
- Traffic surveillance and monitoring
- Intelligent parking management
- Vehicle-to-vehicle communication
- Vehicle-to-infrastructure communication
- Autonomous driving system (ADS)
- Intelligent toll collection system
- Incident/accident detection system
- Smart emergency response system

This list is not inclusive, but many other aspects are included in ITS. If we observe the above-mentioned elements of ITS, we can say that image classification is used in almost every task such as to detect the movement of vehicles; video is used for traffic management, surveillance, and monitoring; images of vehicles or registration plates can be used; for a parking management system, images of empty spaces can be used.

Also, it can be noted that in all these elements, we can see that there is a need for communication between different nodes which includes the transmission and reception of the images or video.

7.3.1 Importance of Semantic Segmentation in Communication Systems for ITS

As mentioned earlier, in the intelligent transportation system (ITS), data transmission is a vital block that is responsible for transmitting and receiving different types of data such as images, text, video, or audio. But mostly, visual information is required to be exchanged. Moreover, this data exchange should be fast enough because safety-critical decisions are to be made based on this information. There are two types of communication

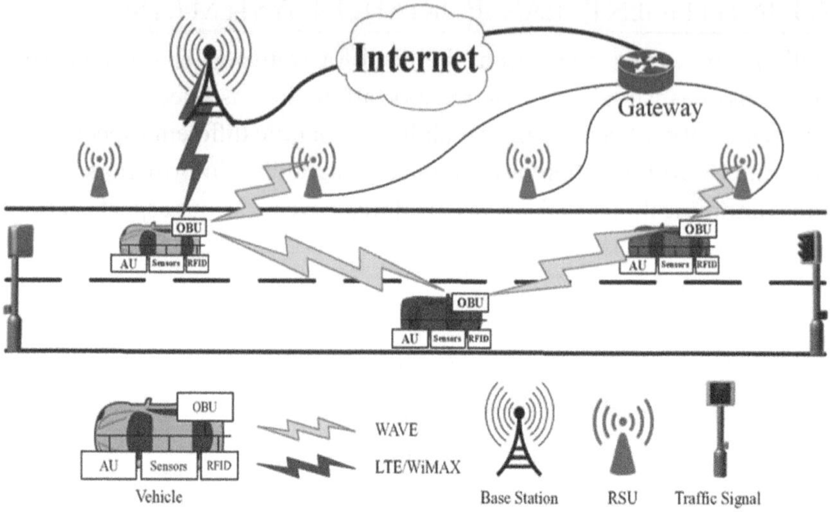

FIGURE 7.10 Internet of vehicles showing vehicle-to-vehicle (V2V) and vehicle-to-infrastructure (V2I) communication [9].

in ITS, one is wired and the other one is wireless. As an example of wired communication, we can discuss an autonomous vehicle in which there is a huge amount of data that is to be transferred between sensors, controllers, and actuators. This is called intra-vehicle communication which includes interaction between human/vehicle and between vehicle and a thing. This communication should be fast, robust, and accurate as the safety of the vehicle is dependent on it. The example of wireless communication involves the exchange of data between different vehicles on the road which provide information to each other for increasing the decision capabilities of the ADS of all the autonomous cars on the road. This communication between the vehicles or from a vehicle to an infrastructure is also called IoV or inter-vehicle communication. Figure 7.10 shows the different modes of communication such as intra-vehicle, inter-vehicle (V2V), and vehicle-to-infrastructure (V2I).

Regardless of which type of communication model is used, the data communication system should have lower latency and high throughput and should utilize less bandwidth. In order to understand the importance of semantics in communication, first understand the meaning of semantics, which is context or meaning. As a human being, if we see an image, we not only know about all the objects in an image but also understand what is happening in an image. To understand this concept, consider Figure 7.11.

FIGURE 7.11 An image showing different people and objects. There is high-level information in the image such as there are people and some objects and then there is low-level information such as people are in good moods or the meeting is successful [10].

Without context, we can see that this image has a certain number of people; there are objects such as a table, a laptop, and a notice board. The understanding without a context or meaning is not sufficient to grasp the complete information in the image. So, if we see the image with context, we will come to the conclusion that there is a meeting going on in the image, and probably a deal has been made or two groups have come to an understanding, which is why they are shaking hands and everyone looks happy.

The traditional communication system does not care for the context in an image, it just transmits the information of pixels in the form of 1's and 0's. It transmits the raw data in the form of bits.

But on the way to the receiver, in the channel, if even a small noise distorts the data, the image will not be reconstructed on the receiver side efficiently. But if the information was sent with context or the features of the image that defines the context, then the image would have been reconstructed with high accuracy even with a small level of noise.

Getting the semantics in text, images, audio, or video is challenging but with the rapid advancement in the hardware of communication and AI, it is becoming possible to replace traditional communication with semantic communication.

Following are some of the benefits of using semantic communication which makes it very useful for real-time and safety-critical applications like intelligent transportation systems:

- In semantic communication, the features are extracted, which describe the context of an image. This makes the data less prone to noise, and it is reconstructed easily on the receiver side. Moreover, if the receiver is getting more focused and relevant data, it can efficiently reconstruct the image.
- Only extracted features or regions of interests (ROIs) are transmitted instead of the whole image to save the bandwidth requirement as it is limited in the case of ITS, especially in intra-vehicle communication.
- For applications like traffic surveillance and monitoring, it is efficient to send only segmented portions of the image that are useful for the application. For example, why send the image of a whole car if only the registration plate is required?
- If an image has multiple objects and these are required by different applications, then we need to transmit the whole image. But if these applications do not need the required object or ROI, then the image can be segmented and clusters can be sent according to the required priority of use. For instance, consider a scenario similar to Figure 7.12. Imagine a car's sensor capturing an image of the road that includes multiple objects. These could be a pedestrian directly in front of the car, other vehicles in an adjacent lane, and a left-turn sign located 100 meters away

FIGURE 7.12 Depiction of the scenario in which an autonomous vehicle is moving in an environment with different objects such as trees, pedestrians, signal, and lane markings. The information about different objects can be prioritized if the image is segmented [11].

Now, if the image is segmented, it does not have to send all the objects in the image to the control unit of ADS, but it can send the information related to the pedestrian in front of the car first so that the control can actuate the mechanism to stop the car. Then, it can send the segment of the signal turning yellow so that the ADS can make the decision accordingly. There is no need to send the information related to the cars coming in the adjacent lane at this very moment because cars will be stopped due to the pedestrian and then will remain in that state because of the signal turning red. So, this example clearly shows that prioritizing the segments can save the bandwidth, lower the processing, and lower the latency by removing the communication overhead in this image.

7.4 TECHNIQUES FOR SEMANTIC SEGMENTATION AND SEMANTIC COMMUNICATION

By now, we have understood the concept of semantic segmentation, its importance, and its types. Now, we will discuss some of the techniques used for semantic segmentation starting from more traditional approaches and then some end-to-end methods based on deep learning. Also, we will discuss some complete communication systems based on the semantically segmented data.

7.4.1 Conventional Approaches for Semantic Segmentation

First, the images are passed through some preprocessing layer in which steps like inverse perspective transformation (IPM), image enhancement, and ROI extraction are involved [12]. The next step involves the separation of ROI from the background, which is called detection. For candidate region detection or feature extraction, histograms of gradients (HOGs) are used which represent the features of an object based on the gradients of pixel groups in all directions [13]. Maximally stable extremal regions (MSERs) can also be used for detection purposes [13,14]. Detection is followed by recognition or classification in which extracted features are labeled according to their class. Initially, recognition was done using template matching in which a database of features related to symbols was created, and then, it is compared against the unknown image [15]. As discussed, image classification using the conventional approach involves separate layers and each layer uses different algorithms. As each layer is responsible for solving a particular task, so, when any changes are made in one layer, it is difficult to

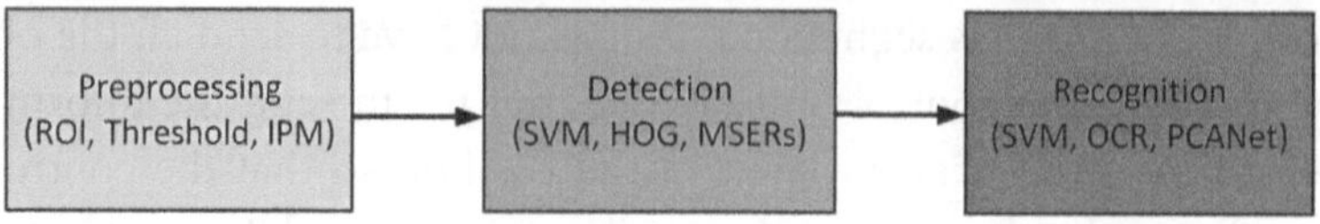

FIGURE 7.13 Flow diagram of a conventional approach for image classification which starts with the preprocessing of the image and then detection is performed using different techniques like support vector machine (SVM), histogram of gradients, etc. and then recognition is performed. All of these stages are separate.

determine how it will impact the whole system. Layers of image recognition using conventional layers are shown in Figure 7.13.

7.4.2 End-to-End Approaches for Semantic Segmentation

End-to-end (E2E) approaches for image classification refer to training a possibly complex learning system by a single model. These models represent the complete target system and bypass the intermediate layers usually present in traditional pipeline designs as discussed earlier. It means that E2E models can predict output directly given the input image. In literature, there are many techniques that are used for semantic segmentation, but we will discuss some of them that are used for the application of ITS. In ITS, the most important task is environment perception which uses the classification of objects in images such as pedestrians, other vehicles, signals, road lanes, and symbolic road markings (SRMs). So, we will discuss mainly those methods which can be used for semantic segmentation.

7.4.3 SegNet – A Deep Convolutional Encoder-Decoder Architecture for Image Segmentation

SegNet [16] consists of a combination of encoder and decoder. This technique is focused on the road scene perception and tested on the CamVid road scene dataset. Consider the architecture of SegNet in Figure 7.14.

The encoder is identical to the convolution layer of VGG16 except that it has no fully connected layer, but the purpose is the same, i.e., the production of feature maps. It is the decoder and its implementation technique that distinguishes SegNet from other semantic segmentation techniques. The decoder recovers the loss of spatial information that is produced due to the pooling layers of the encoder using features from the pooling layers of the encoder; as a result, it enhances the boundary delineation, decreases the number of training parameters, and can be used in any encoder-decoder configuration with few changes. To show the true performance of

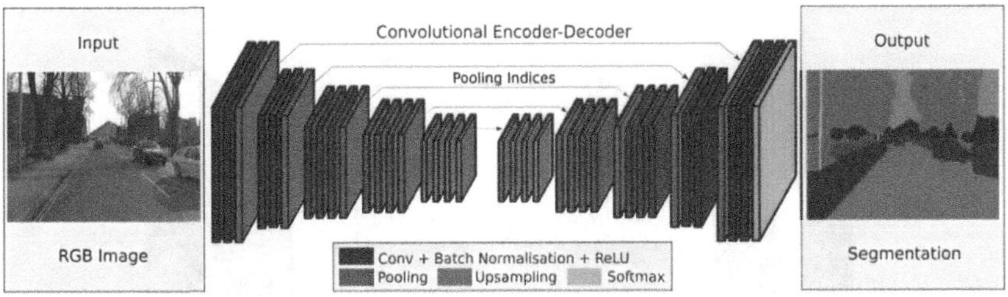

FIGURE 7.14 SegNet architecture for semantic segmentation using encoder-decoder approach [16].

the algorithm, authors have used controlled benchmarking and kept the stochastic gradient descent (SGD) throughout their end-to-end training, but this has increased the model vs optimization issues due to the gradient backpropagation issues and non-convergence nature of optimization for a large number of parameters, thus giving an accuracy of 57.7% and inference time of 60 ms.

7.4.4 ENet: A Deep Neural Network Architecture for Real-time Semantic Segmentation

ENET [17] focuses on the real-time implementation of the semantic segmentation of road scenes for high-speed inference. This technique is tested on a cityscapes dataset, giving an accuracy of 58.3%, a capacity of 79 fps, and an inference time of 13 ms which is improved as compared to SegNet. This technique uses the encoder/decoder technique but without post-processing steps, smaller decoder size, and up-sampling at the decoder using the indices of max pooling elements which makes it faster by minimizing the memory requirements. But to reduce the number of floating-point functions, it is observed that during non-linear operations, inference time increases due to increased memory calls.

7.4.5 FCN: End-to-End Deep Neural Network

Long et al. [18] propose an end-to-end training method for pixel-wise classification keeping the input and output size equal. This enables the algorithm to take the arbitrary size of the input image instead of the fixed size which enhances the accuracy of real-time implementation. Consider the architecture of FCN in Figure 7.15.

An FCN utilizes an encoder and a decoder to produce feature maps retaining spatial information using skip connections passing low-layer

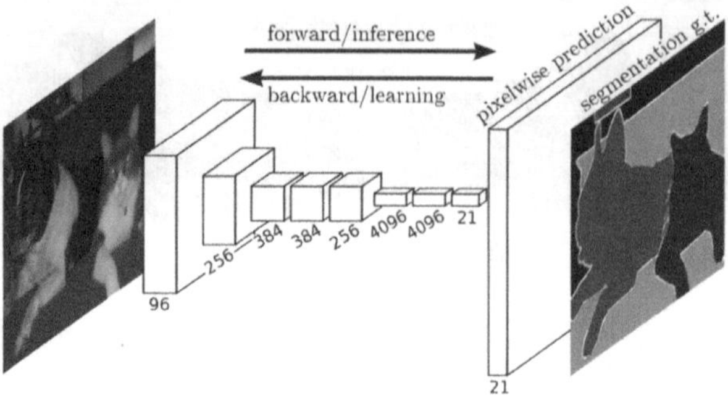

FIGURE 7.15 Architecture for fully convolutional network (FCN) for semantic segmentation [18].

feature information to higher layers. This technique is tested on PASCAL VOC giving the accuracy of 65.3%, 2 fps, and 500 ms inference time. But in VGG16-based FCN, if the number of strides is reduced to get finer information during up-sampling, the size of the fully convolutional filter is increased significantly, which not only increases the computational cost but also makes it difficult to learn such large filters which affect the overall performance of this method. Also, during learning, the problem of gradient vanishing causes a loss in feature information. This technique covers the loss of feature information due to pooling layers but not due to the gradient vanishing which saturates the accuracy [19].

7.4.6 FRNN: Full Resolution of Residual Networks with Skip Connections

The solution of the problem of gradient vanishing is discussed in He et al. [19] which is caused when the number of layers is increased. The proposed method uses residual networks which transfer feature information from lower layers to higher layers. Pohlen et al. [20] has utilized the power of residual networks for the semantic segmentation of road scenes and solves not only the problem of spatial perspective loss but also the problem of gradient vanishing, thus increasing the dense prediction and localization accuracy. This method is called full resolution residual networks (FRNN) and introduces two elements in the pipeline, one method is a residual element that performs residual operations to keep the spatial resolution and the other element is the pooling element, which performs convolutions and pooling to get the feature maps. This algorithm is tested on the cityscapes dataset giving an accuracy of 71.8%. However, this technique is not suitable for real-time applications as it has an inference time of 496 ms.

7.4.7 Dilation 10

A deep neural network introduced by Yu and Koltun [21] employs a modified convolutional architecture that eliminates pooling and subsampling layers, thereby preserving the spatial resolution of the input image in the output. Dilated convolutions are used which aid the exponential growth of receptive fields, avoiding the loss of resolution. This module is used with existing dense prediction algorithms and is tested on many datasets of urban scenes like CamVid, KITTI, and cityscapes datasets with the corresponding accuracies of 65.3%, 59.2%, and 67.1%. This algorithm is not a complete pipeline for semantic segmentation but a single module that can be embedded in existing pipelines; therefore, this solution does not provide end-to-end training.

7.4.8 LDFNet

A deep neural network that increases computational efficiency [22] combines the depth and luminous information along with Red, Green and Blue (RGB) data to increase the computational efficiency by reducing the number of parameters. Table 7.1 shows how much it reduces the number of parameters when compared with other RGB techniques and increases in speed due to a reduction in the number of parameters. Comparison of LDFNet with some RGB techniques shows a substantial drop in the number of parameters, thus increasing the inference speed feature maps with a minimum number of parameters and uses Depth and Luminous (D&Y) encoders to improve the balance between accuracy and speed. This network is tested on the cityscapes dataset with an accuracy of 71.3%

TABLE 7.1 Comparison of LDFNet with Other Neural Networks for Semantic Segmentation

Method	Parameter	Speed (FPS)
Dilation 10	148.8M	0.25
FCN-8s	134.5M	2.0
SegNet	29.5M	16.7
LDFNet	2.31M	18.4

Note: It can be observed that LDFNet has a minimum number of parameters/model size and the highest speed which means that it can process 18.4 images in 1 second.

and a speed of 18.4 fps. This technique is very unique as compared to many existing methods as they only utilize RGB information. Testing this method on cityscapes only limits the performance as this dataset does not have accurate disparity images [23].

Above-discussed techniques are some of the ways to segment the image semantically. Now, let's discuss some of the important methods for the implementation of semantic communication.

7.4.9 Method 1 [24]

A typical communication system has three components.

1. A transmitter
2. Channel
3. Receiver

An image is transmitted by converting the image into a binary stream at the transmitter stage. It is then transmitted over a communication channel which can be wired or wireless. Figure 7.16 shows the architecture for this technique. The receiver then reconstructs the image. As mentioned earlier, this method is more prone to noise, and the image is not reconstructed properly at the receiver. On the other hand, in this method, the elements of communication are also three, but the transmitted semantic segmentation is used for feature extraction, and then, these features are transmitted through the communication channel. Figure 7.16 shows the communication system for transmitting the image using semantic segmentation.

The first part is the transmitter which has a feature extraction block using a semantic segmentation technique, a semantic encoder, and a compression block. For semantic segmentation, a method called U-Net [25] is used. This is a CNN that can perform pixel-wise labeling of objects in an image. It has one encoder and one decoder. Each of these has different layers like convolution, maxpool, and rectified liner unit (ReLU). The encoder downsamples the image and gets the features or the context of different objects in the image. On the other hand, the purpose of the decoder is to up-sample the image and fuse the extracted features to reconstruct the image with pixel-wise labeled information. Our objective is to get the ROI and region of non-interest (RONI) from the image. Therefore, two U-Net models are used. So, the semantic segmentation block gives pixel-wise labeled images, semantic encoder 1 gives ROI, and semantic encoder

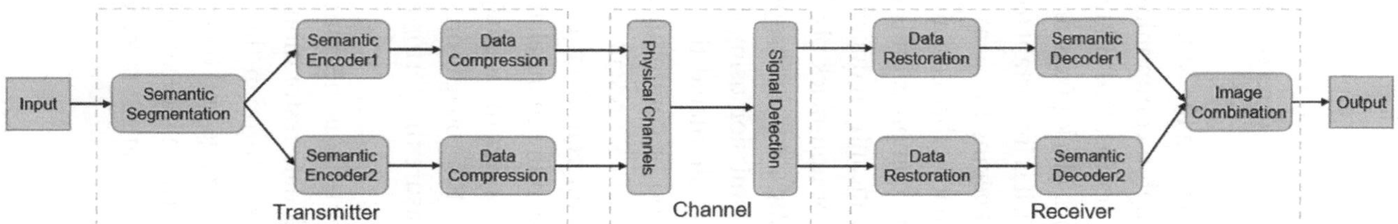

FIGURE 7.16 Architecture for semantic communication in which semantic information is extracted in the transmitter, transferred through the channel after compression, and the decoder is used to reconstruct the image at the receiver [24].

2 gives RONI. At this stage, there is one issue and that is the use of convolution layers. The size of the image reduces when it passes through the convolution layer. In semantic communication, it is important to keep the size of the input and output the same. Therefore, we have used zero-padding to perform the same convolution and it resolves our issue. The semantic decoder is used on the receiver side to reconstruct the image using the information of ROI and RONI.

On the transmitter side, after getting the ROI and RONI, the features are transmitted to the receiver through a communication channel. Data from encoder 1 and encoder 2 will be transmitted simultaneously, but the data is sparse and most of the region has black pixels or 0 values which increases the communication overhead. So, it is important to compress the data before transmitting it. The padding for keeping the size of the image the same after convolution produces a sparse matrix that contains mostly black pixels. To compress this data, sparse matrix compression is used in this method. The communication overhead is significantly reduced using this approach as it will select only important pixels or non-black pixels for transmission. Here, we can see the benefits of semantic communication discussed previously.

Now, the data is ready for transmission using a communication channel. Researchers in this method have simulated the channel instead of using a physical channel. In simulation, a communication channel is perfect because there are no real-time issues present in the channel. The major element that makes simulation different from real-time channels is noise. So, to observe the effects of noise in the channel, this method uses "additive white Gaussian noise" or AWGN. Also, the Rayleigh fading channel is used to simulate the effects of multipath propagation on the signals. Signal-to-noise ratio (SNR) is used to measure the performance of the communication channel.

The receiver also has three elements as the transmitter, but the purpose is different. When data is received, data restoration is performed for decompression. After that, data is passed through the semantic decoder, which reconstructs the two parts of the image and then these two parts are fused to make the final image. The performance of this algorithm is measured through SNR and compression ratio at the receiver side.

7.4.10 Method 2 [26]

In the previous method, only high-level information was transmitted. However, an image can have multi-level information or features.

A high-level feature extractor gives the overall context or meaning of an image while a low-level feature extractor investigates more details. For example, in Figure 7.10, a high-level feature will show that a meeting is going on in the image or any text, but low-level details will include that people are happy, the colors of clothes, or the texture of clothes.

The method under discussion is also used for transmitting images using multi-level semantic information for efficient bandwidth utilization. This is done mainly by reducing the communication overhead, but reducing the information can cause loss of information as well. That is why this method is designed to extract features from an image on multiple levels. More specifically, it extracts high-level features such as text and segment-based meaning and low-level features such as shapes, texture, or color information in the image. The multi-level semantic communication can be seen in Figure 7.17.

This semantic communication system also uses an encoder and decoder approach. The encoder extracts the multi-level features from an input image (S), encodes the data into symbols, and then transmits it over a noisy channel. The decoder, at the receiver side, extracts features from the symbols and then reconstructs the image (S') using this information. The method starts by normalizing the image pixel values between 0 and 1. Three different neural networks are used for extracting different levels of features. For getting semantic information in the form of text, CaptionNet [27] based on ResNet-152 [28] and Long Short-Term Memory (LSTM) [29] is used. This network produces captions or text embedding based on the high-level features in the image. The segmentation-based features are extracted using the pyramid scene parsing network (PSPNet) [30]. For extracting low-level information, the concatenation technique is used to combine the normalized image and segment-based features. The encoder

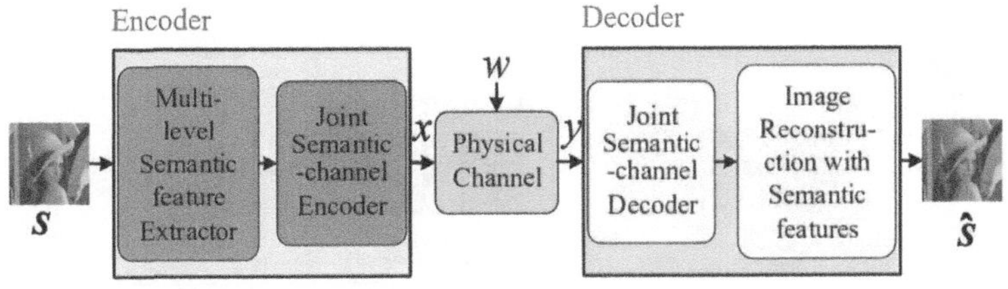

FIGURE 7.17 Semantic communication method which is capable of extracting high-level and low-level semantics from an image at transmitter and then reconstructed at the receiver [26].

converts this multi-level information and transmits it over a communication channel. The channel has AWGN noise like method one. Once the data is received on the receiver, a decoder is used to remove the noise and extract the multi-level feature information. This data is then fused to reconstruct the images.

7.4.11 Method 3 [31]

In the previous two methods, it is mentioned that the reconstruction of images transmitted using semantic segmentation is of good quality. But in today's world of image sharing, especially on mobile networks, a communication system demands transmission, reception, and reconstruction of high-quality images. This method not only implements semantic communication but also ensures that high-quality images are reconstructed at the receiver end. This method applies end-to-end semantic communication which consists of a transmitter, channel, and receiver. It also uses an encoder and decoder approach. This system is GAN (generative adversarial network) which produces or generates new images based on the provided data. It is used at the receiver end. The basic functionality of this GAN-based semantic communication system is shown in Figure 7.18.

On the transmitter side, a semantic segmentation map of an image is made using a pre-trained network on the common objects in context (COCO) dataset [32]. Note that, the GAN at the receiver is also pre-trained on the same dataset. This means that the dataset is the knowledge base for the transmitter and receiver. After converting the image into a segmented map, channel encoding is applied using a binary phase-shift keying

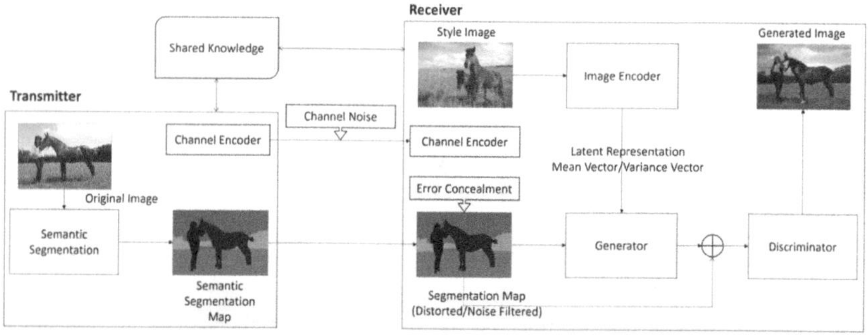

FIGURE 7.18 Semantic communication system in which both transmitter and receiver are trained with same dataset. GAN is used at the receiver for the efficient reconstruction of the image [31].

(BPSK) encoding scheme. This data is transmitted through a noisy channel (AWGN) and is detected on the receiver. The receiver first performs the channel decoding. It then inputs the data into GAN, which uses the shared knowledge base and received data to construct the real-time high-quality image. The compression ratio of 20 is achieved using this technique which shows a good performance.

7.5 CONCLUSION

Transmitting an image is costly in terms of bandwidth. In many systems, such as ADS, not all the information from an image is required. Sending a complete image has communication overhead which increases the bandwidth utilization and latency. Moreover, the images in the communication systems are transmitted in the form of bits which represent all the pixels in an image. This information is prone to noise and reconstruction of images as the receiver is not accurate. In this chapter, we have presented a technique called semantic segmentation which segments the objects in an image. Using this method, the context of the picture can also be understood. Using this technique, only required information can be transmitted with reduced communication overhead, and also the information about context or semantics can help in the efficient reconstruction of an image on the receiver.

REFERENCES

[1] Lin, Tsung-Yi, Michael Maire, Serge Belongie, James Hays, Pietro Perona, Deva Ramanan, Piotr Dollár, and C. Lawrence Zitnick. "Microsoft coco: Common objects in context." In *Computer Vision–ECCV 2014: 13th European Conference, Zurich, Switzerland, September 6–12, 2014, Proceedings, Part V 13*, pp. 740–755. Springer International Publishing, 2014.

[2] Mustaffa, M. T., N. M. Z. Hashim, N. M. Saad, N. A. A. Hadi, A. Salleh, and A. S. Ja'afar. "Biomedical signal and image processing in Matlab." *International Journal for Advance Research in Engineering and Technology*, 3, no. IX (2001): 29–32.

[3] Ma, Wei-Ying, and Bangalore S. Manjunath. "EdgeFlow: A technique for boundary detection and image segmentation." *IEEE Transactions on Image Processing*, 9, no. 8 (2000): 1375–1388.

[4] Mittal, Himanshu, Avinash Chandra Pandey, Mukesh Saraswat, Sumit Kumar, Raju Pal, and Garv Modwel. "A comprehensive survey of image segmentation: Clustering methods, performance parameters, and benchmark datasets." *Multimedia Tools and Applications,* 81, no. 24 (2021): 1–26.

[5] Felzenszwalb, Pedro F., and Daniel P. Huttenlocher. "Efficient graph-based image segmentation." *International Journal of Computer Vision,* 59 (2004): 167–181.

[6] Guo, Yanming, Yu Liu, Theodoros Georgiou, and Michael S. Lew. "A review of semantic segmentation using deep neural networks." *International Journal of Multimedia Information Retrieval,* 7 (2018): 87–93.

[7] Romera-Paredes, Bernardino, and Philip Hilaire Sean Torr. "Recurrent instance segmentation." In *Computer Vision–ECCV 2016: 14th European Conference, Amsterdam, The Netherlands, October 11–14, 2016, Proceedings, Part VI 14,* pp. 312–329. Springer International Publishing, 2016.

[8] Ghosh, Swarnendu, Nibaran Das, Ishita Das, and Ujjwal Maulik. "Understanding deep learning techniques for image segmentation." *ACM Computing Surveys (CSUR),* 52, no. 4 (2019): 1–35.

[9] Cheng, JiuJun, JunLu Cheng, MengChu Zhou, FuQiang Liu, ShangCe Gao, and Cong Liu. "Routing in internet of vehicles: A review." *IEEE Transactions on Intelligent Transportation Systems,* 16, no. 5 (2015): 2339–2352.

[10] Pexels. www.pexels.com/search/group%20of%20people/

[11] Graziano, Leonardo, "Autonomi autonomous mobility interface," Online, 2014, 2018-02-3. [Online]. Available: https://vimeo.com/99160686

[12] Bailo, Oleksandr, Seokju Lee, Francois Rameau, Jae Shin Yoon, and In So Kweon. "Robust road marking detection and recognition using density-based grouping and machine learning techniques." In *2017 IEEE Winter Conference on Applications of Computer Vision (WACV), Santa Rosa, CA, March 24–31 2017,* pp. 760–768. IEEE, 2017.

[13] Dalal, Navneet, and Bill Triggs. "Histograms of oriented gradients for human detection." In *2005 IEEE Computer Society Conference on Computer Vision and Pattern Recognition (CVPR'05), San Diego, CA, June 20–26 2005,* vol. 1, pp. 886–893. IEEE, 2005.

[14] Greenhalgh, Jack, and Majid Mirmehdi. "Automatic detection and recognition of symbols and text on the road surface." In *Pattern Recognition: Applications and Methods: 4th International Conference, ICPRAM 2015, Lisbon, Portugal, January 10–12, 2015, Revised Selected Papers 4*, pp. 124–140. Springer International Publishing, 2015.

[15] Amayo, Paul, Tom Bruls, and Paul Newman. "Semantic classification of road markings from geometric primitives." In *2018 21st International Conference on Intelligent Transportation Systems (ITSC), Maui, HI, November 4–7, 2018*, pp. 387–393. IEEE, 2018.

[16] Badrinarayanan, Vijay, Alex Kendall, and Roberto Cipolla. "Segnet: A deep convolutional encoder-decoder architecture for image segmentation." *IEEE Transactions on Pattern Analysis and Machine Intelligence*, 39, no. 12 (2017): 2481–2495.

[17] Paszke, Adam, Abhishek Chaurasia, Sangpil Kim, and Eugenio Culurciello. "Enet: A deep neural network architecture for real-time semantic segmentation." *arXiv preprint arXiv:1606.02147* (2016).

[18] Long, Jonathan, Evan Shelhamer, and Trevor Darrell. "Fully convolutional networks for semantic segmentation." In *Proceedings of the IEEE Conference on Computer Vision and Pattern Recognition, Boston, MA, June 7–12, 2015*, pp. 3431–3440. 2015.

[19] He, Kaiming, Xiangyu Zhang, Shaoqing Ren, and Jian Sun. "Deep residual learning for image recognition." In *Proceedings of the IEEE Conference on Computer Vision and Pattern Recognition, Las Vegas, NV, June 27–30, 2016*, pp. 770–778. 2016.

[20] Pohlen, Tobias, Alexander Hermans, Markus Mathias, and Bastian Leibe. "Full-resolution residual networks for semantic segmentation in street scenes." In *Proceedings of the IEEE Conference on Computer Vision and Pattern Recognition, Honolulu, HI, July 21–26, 2017*, pp. 4151–4160. 2017.

[21] Yu, Fisher, and Vladlen Koltun. "Multi-scale context aggregation by dilated convolutions." *arXiv preprint arXiv:1511.07122* (2015).

[22] Hung, Shang-Wei, Shao-Yuan Lo, and Hsueh-Ming Hang. "Incorporating luminance, depth and color information by a fusion-based network for semantic segmentation." In *2019 IEEE International conference on image processing (ICIP), Taipei, Taiwan, September 22–25, 2019*, pp. 2374–2378. IEEE, 2019.

[23] Seichter, Daniel, Mona Köhler, Benjamin Lewandowski, Tim Wengefeld, and Horst-Michael Gross. "Efficient RGB-D semantic segmentation for indoor scene analysis." In *2021 IEEE International Conference on Robotics and Automation (ICRA), Xi'an, China, May 30 to June 5, 2021*, pp. 13525–13531. IEEE, 2021.

[24] Wu, Jiale, Celimuge Wu, Yangfei Lin, Tsutomu Yoshinaga, Lei Zhong, Xianfu Chen, and Yusheng Ji. "Semantic segmentation-based semantic communication system for image transmission." *Digital Communications and Networks,* 10, no. 3 (2023): 519–527.

[25] Ronneberger, Olaf, Philipp Fischer, and Thomas Brox. "U-Net: Convolutional networks for biomedical image segmentation." In *Medical Image Computing and Computer-Assisted Intervention–MICCAI 2015: 18th International Conference, Munich, Germany, October 5–9, 2015, Proceedings, Part III 18*, pp. 234–241. Springer International Publishing, 2015.

[26] Zhang, Zhenguo, Qianqian Yang, Shibo He, Mingyang Sun, and Jiming Chen. "Wireless transmission of images with the assistance of multi-level semantic information." In *2022 International Symposium on Wireless Communication Systems (ISWCS), Hangzhou, China, October 19–22, 2022*, pp. 1–6. IEEE, 2022.

[27] Yang, Longyu, Hanli Wang, Pengjie Tang, and Qinyu Li. "CaptionNet: A tailor-made recurrent neural network for generating image descriptions." *IEEE Transactions on Multimedia,* 23 (2020): 835–845.

[28] Athisayamani, Suganya, Robert Singh Antonyswamy, Velliangiri Sarveshwaran, Meshari Almeshari, Yasser Alzamil, and Vinayakumar Ravi. "Feature extraction using a residual deep convolutional neural network (ResNet-152) and optimized feature dimension reduction for MRI brain tumor classification." *Diagnostics,* 13, no. 4 (2023): 668.

[29] Sunny, Md Arif Istiake, Mirza Mohd Shahriar Maswood, and Abdullah G. Alharbi. "Deep learning-based stock price prediction using LSTM and bi-directional LSTM model." In *2020 2nd Novel Intelligent and Leading Emerging Sciences Conference (NILES), Giza, Egypt, October 24–26, 2020*, pp. 87–92. IEEE, 2020.

[30] Yan, Lingfei, Dawei Liu, Qi Xiang, Yang Luo, Tao Wang, Dali Wu, Haiping Chen, Yu Zhang, and Qing Li. "PSP net-based automatic segmentation network model for prostate magnetic resonance

imaging." *Computer Methods and Programs in Biomedicine,* 207 (2021): 106211.

[31] Lokumarambage, Maheshi U., Vishnu Sai Sankeerth Gowrisetty, Hossein Rezaei, Thushan Sivalingam, Nandana Rajatheva, and Anil Fernando. "Wireless end-to-end image transmission system using semantic communications." *IEEE Access,* 11 (2023): 37149–37163.

[32] Lin, Tsung-Yi, Michael Maire, Serge Belongie, James Hays, Pietro Perona, Deva Ramanan, Piotr Dollár, and Lawrence Zitnick. "Microsoft coco: Common objects in context." In *Computer Vision–ECCV 2014: 13th European Conference, Zurich, Switzerland, September 6–12, 2014, Proceedings, Part V 13*, pp. 740–755. Springer International Publishing, 2014.

[33] Minaee, Shervin, Yuri Boykov, Fatih Porikli, Antonio Plaza, Nasser Kehtarnavaz, and Demetri Terzopoulos. "Image segmentation using deep learning: A survey." *IEEE Transactions on Pattern Analysis and Machine Intelligence,* 44, no. 7 (2021): 3523–3542.

CHAPTER 8

Illuminate the Pathways of Moth Flame Optimization for Global Optimization

Applications in Signal Processing and AI for Optical Communication Systems

Srushti M P, Saanvi Ravi, Pushkar V, Samjith T S, Samarth V Gangadikar, and Divya C D

8.1 INTRODUCTION

Navigating the landscape of optimization algorithms: As we stand on the brink of an intriguing exploration, we are drawn to the profound impact of optimization algorithms that intricately shape diverse domains. These algorithms possess a unique allure, unlocking solutions to intricate problems and sparking the pursuit of innovative metaheuristics. Amidst the expansive array of problem-solving techniques, our attention zeroes in on the enigmatic moth flame optimization (MFO) algorithm, promising novel perspectives and transformative insights into global optimization

DOI: 10.1201/9781003472506-8

within the context of signal processing and artificial intelligence (AI) in optical communication systems.

The Enchanting Genesis of MFO: Within this section, we immerse ourselves in the captivating tale of nature's wisdom—the genesis of MFO. The narrative unfolds gracefully, tracing the heuristic's origins to the intriguing behavior of moths, akin to intrepid explorers irresistibly drawn to the enchanting glow of flames. From this inspiration, the architects of MFO harnessed nature's ingenuity to craft an algorithm mirroring the spirit of these winged creatures. Venturing into the core of MFO, we encounter its elemental components: the fluttering moth population, symbolic of exploration and diversity; the magnetic pull of light intensity, representing attraction and convergence; the mesmerizing dance of flames, embodying refining and optimization; and the soaring flight equations leading to optimal solutions, mirroring the elegant trajectory of moths in pursuit of light. MFO's biologically inspired brilliance beckons us to explore the depths of its efficiency and effectiveness.

Embarking on an Intellectual Odyssey: We boldly declare our noble quest—an intellectual odyssey transcending conventional optimization, venturing into uncharted territories. Our minds, illuminated by the incandescent spark of inquiry, propel us on a relentless journey to fathom the essence of the MFO algorithm. What distinguishes MFO from its algorithmic counterparts? How does it ingeniously weave together nature's stratagems to reshape global optimization, particularly within the realms of signal processing and AI in optical communication systems? These persistent questions guide our exploration, casting light upon our path as we navigate the intricate labyrinth of knowledge. With each step, we unveil the cryptic layers enshrouding MFO, progressively unearthing the invaluable treasures it conceals in its quest to conquer optimization challenges.

In this academic voyage, our exploration extends beyond conventional boundaries, forging a symbiotic link between MFO's principles and the intricate dynamics of optical and wireless communication systems. We meticulously dissect how MFO's innate wisdom finds resonance within these domains, harnessing its adaptive brilliance to revolutionize signal transmission, channel allocation, and spectral efficiency. By interrogating MFO's role in the communication spectrum, we illuminate novel avenues for enhancing data throughput, minimizing interference, and amplifying connectivity.

As we navigate this uncharted terrain, our unwavering determination kindles a beacon of insight, guiding us to unravel the enigma of MFO's design. Through this synergy of exploration and illumination, we chart a course that promises to reshape optimization paradigms within the realms of signal processing and AI in optical communication systems, propelling us toward a future where the convergence of nature's elegance and cutting-edge technology ignites unprecedented frontiers of connectivity and efficiency.

8.2 LITERATURE REVIEW

This section serves as a captivating journey through the diverse landscape of optimization techniques, offering a panoramic view of problem-solving approaches that have shaped the field of optimization over the years. We peer into a kaleidoscope of methodologies, ranging from classic deterministic algorithms to cutting-edge metaheuristics, each with its unique strengths and limitations. As we traverse this rich tapestry of optimization approaches, we are drawn toward the radiant glow of the MFO algorithm, standing out as a gem that beckons us with promises of novel perspectives and transformative insights into global optimization.

The symphony of optimization techniques begins with the classic elegance of deterministic algorithms, such as gradient-based methods and Newton's method. These algorithms showcase precision and efficiency in solving well-defined problems with smooth objective functions. However, their reliance on gradient information makes them susceptible to getting trapped in local optima, limiting their effectiveness in handling complex, multimodal, and noisy optimization landscapes.

As our gaze shifts, we encounter the allure of evolutionary algorithms, such as genetic algorithms (GA) and differential evolution (DE) [1]. These algorithms mimic the principles of natural selection and evolution, exploring diverse regions of the search space and exhibiting remarkable robustness in handling complex, multimodal problems. Nevertheless, the computational burden of evaluating a large population of solutions and the challenge of balancing exploration and exploitation remain noteworthy considerations.

The symphony crescendos as we immerse ourselves in the world of metaheuristics, where innovation flourishes through a delicate balance of exploration and exploitation. Particle swarm optimization (PSO) and ant colony optimization (ACO) [2,3] showcase the power of collective intelligence, inspired by the foraging behavior of birds and ants. The emergent collaboration in these algorithms propels them toward promising solutions. However, the lack of global convergence guarantees and the vulnerability to premature convergence warrant careful scrutiny.

As the symphony reaches its zenith, we find ourselves captivated by the mesmerizing dance of swarm intelligence, where the harmony of exploration and exploitation resounds with profound implications. The charm of the firefly algorithm (FA) [4] lies in its flickering glow, where fireflies communicate through light signals to attract each other, reflecting the essence of attraction in optimization. The harmony of harmony search enchants us [5], as it draws inspiration from the improvisational nature of musical harmony, where solutions harmoniously cooperate to reach optimal states. However, these algorithms, while captivating, also have their limitations, such as the need for fine-tuning parameters and vulnerability to premature convergence in high-dimensional problems.

Amidst this symphony of optimization strategies, the radiant glow of moth flame optimization captures our attention [6]. MFO's allure emanates from its ingenious combination of moth-inspired strategies, accentuated by the ethereal dance of flames. Drawing inspiration from the moths' unyielding attraction to luminous sources, MFO fuses nature's brilliance with human ingenuity, creating an algorithm that explores the search space, converges on optimal solutions, and exhibits remarkable robustness in handling various optimization challenges.

Our journey through the literature paves the way to comprehend MFO's uniqueness and sparks the curiosity to explore its potential in solving real-world optimization problems. As we approach MFO, we are inspired to embrace the elegance of nature's strategies and witness how this heuristic illuminates the paths to global optimization triumph [7], transcending the confines of conventional approaches. With a deeper appreciation of the optimization landscape and the radiant brilliance of MFO, we step forward into the heart of our research, ready to explore the captivating depths of this nature-inspired algorithm.

8.3 METHODOLOGY

8.3.1 Optimizing Signal Processing and AI in Optical Communication Systems: A Modified Moth Flame Optimization (MFO) Algorithm

In the MFO algorithm, several fundamental concepts are utilized to mimic the behavior of moths and optimize solutions. Let's explain these concepts:

1. **Moth Population (*M*):**
The moth population refers to a collection of moths represented by an $N \times D$ matrix called *M*, where *N* is the number of moths, and *D* is the dimensionality of the solution vector that each moth represents. These moths are the search agents exploring the solution space to find optimal solutions to the given optimization problem.
2. **Flame Matrix (*F*):**
The flame matrix *F* is also an $N \times D$ matrix, the same size as *M*, used to approximate the best solutions found so far. It represents the positions of promising solutions discovered during the optimization process. In the context of MFO, "flames" act as beacons or attractive light sources that moths attempt to approach during the search.
3. **Flame Intensity:**
The flame intensity represents the quality or fitness value of the corresponding solutions in the flame matrix *F*. Each flame has an associated fitness value, which quantifies how good the solution represented by that flame is with respect to the optimization problem. The higher the flame intensity, the better the solution is considered.
4. **Attraction Behavior:**
The central concept of MFO is the attraction behavior of moths toward flames (optimal solutions). Inspired by the phototaxis behavior of moths moving toward light, the moths in MFO attempt to approach the flames in the flame matrix *F*. The closer a moth gets to a flame, the more it is influenced by the flame's information, guiding it toward potentially better solutions.
5. **Logarithmic Spiral Movement:**
To model the movement of moths toward flames, the MFO algorithm uses a logarithmic spiral formula. This formula describes the trajectory of a moth as it approaches the corresponding flame. The logarithmic spiral movement pattern helps the moths converge toward flames while maintaining a balance between exploration and exploitation.
6. **Exploration and Exploitation:**
Exploration refers to the process of searching the solution space widely to discover new and unexplored regions. On the other hand,

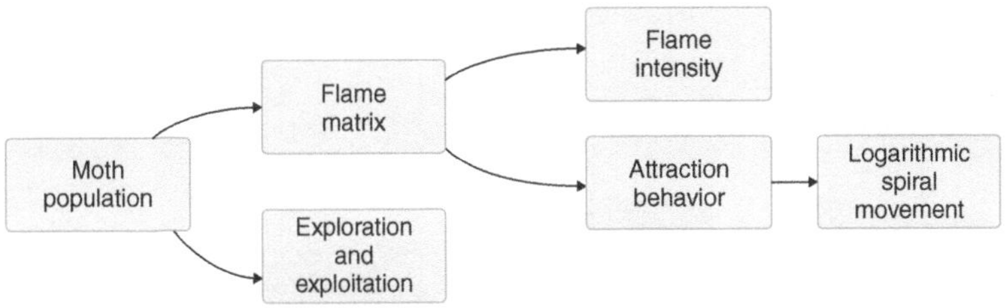

FIGURE 8.1 The MFO algorithm flow diagram.

exploitation involves intensively focusing on already promising regions to find better solutions. MFO balances exploration and exploitation by adjusting the number of flames over iterations and guiding moths toward flames in a controlled manner. The flow diagram of the MFO algorithm is shown in Figure 8.1.

8.3.1.1 Proposed Algorithm

The flow diagram of the proposed algorithm is shown in Figure 8.2, and proceed according to the following steps:

Step 1. Initialization:

- Set the parameters N (number of moths), D (dimensionality of the solution vector), max_iterations (maximum number of iterations), max_flames (maximum number of flames), and b (user parameter).
- Initialize the moth population matrix M with N moths, each with a randomly generated solution vector.
- Initialize the flame matrix F and the corresponding fitness values vector OF with the best solutions found so far. Initially, set F and OF as empty.

Step 2. Optimization Loop:

- Repeat for each iteration l from 1 to max_iterations:
- Compute the fitness values for each moth in M and store them in vector OM.
- Update the flame matrix F and its corresponding fitness values vector OF:

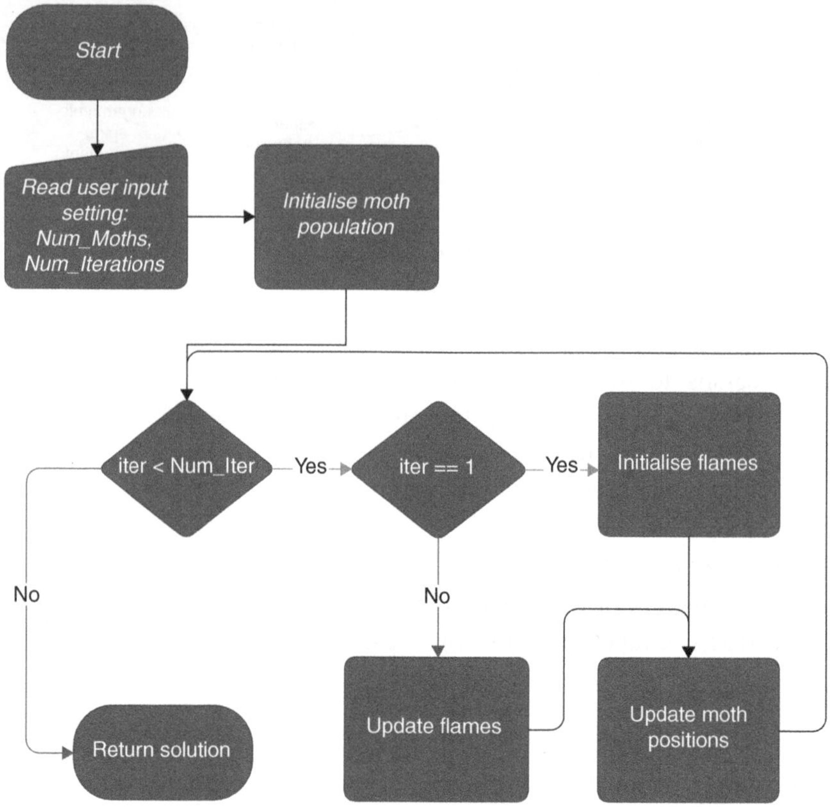

FIGURE 8.2 Flowchart of the proposed algorithm.

- For each moth in M:
- If F is not full (less than max_flames), add the current moth and its fitness value to F and OF, respectively.
- If F is full, compare the fitness value of the current moth (OM[i]) with the fitness value of the worst flame (OF[worst_flame_index]). If OM[i] is better, replace the worst flame with the current moth in F and update OF accordingly.
- Update the number of flames (n) for the next iteration using equation (3): n = max_flames – (max_flames – 1) * l/max_iterations
- Update the moths in relation to the best corresponding flames in F:
- For each moth in M:

- Choose the best corresponding flame in F based on fitness values similarity or other criteria.
- Update the position of the moth using the logarithmic spiral formula: x_new = x_old + exp(−b * t) * cos(2 * pi * t) * (x_old − x_best_flame) y_new = y_old + exp(−b * t) * sin(2 * pi * t) * (y_old − y_best_flame)

  (where (x_old, y_old) is the current position of the moth, (x_best_flame, y_best_flame) is the position of the best corresponding flame, and t is a random variable that decreases the range of the spiral over time).

Step 3. Convergence Check:

- Check if the termination criteria are met (e.g., the solution reaches an acceptable fitness level, or the maximum number of iterations is reached). If yes, stop the optimization process; otherwise, continue to the next iteration.

Step 4. Output

- The optimal solution vector is found in the flame matrix F.

8.3.1.2 Key Components:

The MFO algorithm is a nature-inspired metaheuristic optimization algorithm that aims to find optimal solutions to optimization problems. Comparing the updated version provided with the existing version mentioned in the initial description, here are some potential advantages and improvements:

1. **Enhanced Exploration and Exploitation:** The updated version introduces a dynamic adaptation of the number of flames over iterations. By gradually decreasing the number of flames, the algorithm shifts its focus from exploration (using more flames) to exploitation (using fewer flames). This promotes better balance between

exploring the search space and exploiting promising regions, which can lead to better convergence to global optima.

2. **Improved Convergence:** The use of the logarithmic spiral formula for updating the moths' positions allows for more effective exploration near the flames and convergence toward the best solutions. The spiral-based movement patterns enable moths to follow a smooth trajectory toward the flames, which can enhance convergence speed.
3. **Automatic Flame Management:** The update for the number of flames based on the current iteration allows the algorithm to adaptively adjust its population size. This feature prevents the algorithm from being trapped in local optima by reducing the number of flames over time and focusing on the most promising regions.
4. **Efficient Memory of Best Solutions:** The algorithm maintains a flame matrix F to store the best solutions found so far. This memory mechanism helps in retaining promising solutions across iterations and facilitates better exploitation of valuable information.
5. **Tunable Parameters:** The introduction of user-defined parameters (e.g., b) in the logarithmic spiral formula provides greater flexibility to fine-tune the algorithm's behavior according to the problem characteristics.
6. **Simplicity and Compactness:** The algorithm is relatively compact and straightforward to implement compared to some complex metaheuristic algorithms, making it easy to understand and use.

However, it's important to note that the effectiveness of the updated version of the MFO algorithm may still heavily depend on various factors, such as problem domain, parameter tuning, and the nature of the optimization problem itself. There is no one-size-fits-all optimization algorithm, and the performance of MFO should be evaluated on a case-by-case basis or compared against other state-of-the-art optimization methods to determine its superiority for specific problem instances. Additionally, performance comparisons should consider factors such as convergence speed, solution quality, and computational efficiency.

8.3.1.3 Pseudocode

The pseudocode of the proposed approach is provided below:

```
Function MothFlameOptimization(N, D, max_iterations, max_flames, b):
   Initialize moth population matrix M with N moths, each with a
      randomly generated solution vector
   Initialize flame matrix F and its corresponding fitness values vector OF
      as empty

   For l = 1 to max_iterations:
      // Step 1: Update Flames
      Compute the fitness values for each moth in M and store them in
         vector OM
     If F is not full (less than max_flames):
        Add the best moth and its fitness value to F and OF
     Else:
         Find the index of the worst flame in F (worst_flame_index)
     For each moth i in M:
            If OM[i] is better than OF[worst_flame_index]:
               Replace the worst flame with moth i and update OF
                  accordingly

      // Step 2: Update the number of flames        n = max_flames -
         ((max_flames - 1) * l) / max_iterations

      Step 3: Update Moths
      For each moth i in M:
            Select the best corresponding flame in F based on fitness
         value similarity or other criteria
            Generate a random value t between 0 and 1
         For d = 1 to D:
               Update the position of the moth's d-th dimension using the
         logarithmic spiral formula:
               x_new = x_old + exp(-b * t) * cos(2 * pi * t) * (x_old - x_
         best_flame[d])              y_new = y_old + exp(-b * t) * sin(2 *
         pi * t) * (y_old - y_best_flame[d])          Update the solution
         vector of moth i using the new position (x_new, y_new)

    Return the best solution vector found in the flame matrix F
End Function
```

8.3.1.4 Parameter Settings and Dynamic Adaptation

Static Parameter Settings:

Traditionally, MFO has relied upon static parameter settings, wherein the key algorithmic parameters, such as the population size, the maximum flight distance, and the attraction factors, remain steadfast throughout the optimization endeavor.

Dynamic Parameter Integration:
To infuse MFO with an added edge and elevate its performance, researchers have begun to explore dynamic parameter adaptation strategies. By infusing the algorithm with dynamic parameters, capable of adapting in real time based on the optimization progress and problem characteristics, the optimization process gains newfound agility, bolstering convergence speed and augmenting its robustness in conquering complex optimization landscapes.

8.3.1.5 Performance Evaluation

8.3.1.5.1 Benchmark Functions To cast an incisive glance upon MFO's mettle, a set of painstakingly selected benchmark functions, showcasing an assortment of unimodal and multimodal characteristics, are employed. These benchmark functions serve as a crucible to scrutinize the algorithm's capability to unearth the global optima amidst diverse landscapes.

8.3.1.5.2 Experimental Setup The experimental setup entails subjecting MFO and other comparative algorithms to a grueling examination of the chosen benchmark functions. This examination entails an iterative run for a specified number of generations, with termination criteria dictating halting conditions based on a predetermined maximum generation limit or a threshold for satisfactory solution quality. To ascertain the robustness of the findings, statistical analysis methods, such as mean and standard deviation, are harnessed to draw captivating insights from the empirical results.

8.3.1.5.3 Comparative Analysis The results of these carefully orchestrated experiments are unveiled, meticulously elucidating MFO's prowess in comparison to other optimization algorithms. The performance metrics, captivatingly illustrating convergence speed, solution accuracy, and tenacity, breathe life into the comparative analysis, ultimately leading to an enchanting revelation of MFO's edge over its peers.

The proposed approach benefits from the exploration aspect as shown in Figure 8.3.

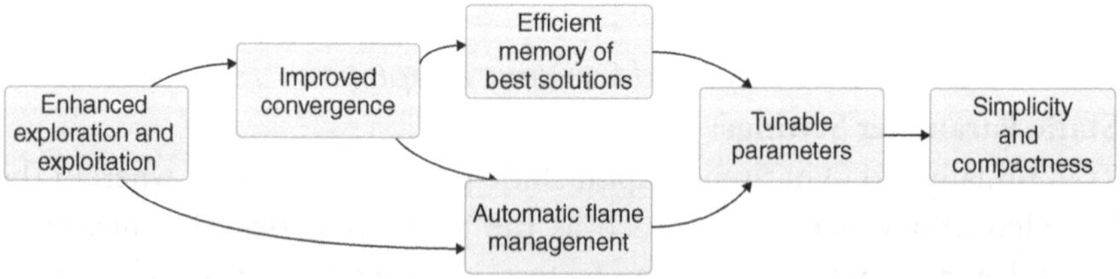

FIGURE 8.3 Exploration diagram.

8.4 APPLICATIONS OF MFO

8.4.1 Real-Life Problem-Solving

The enchanting allure of MFO extends well beyond theoretical conjectures, as it has already found a potent place in real-world optimization problems. These captivating applications span diverse domains, transcending the boundaries of machine learning, engineering, and finance. MFO has adorned itself with the mantle of a potent problem.

8.4.2 Application of MFO in Disaster Response Planning

8.4.2.1 Problem Statement

Disaster response planning involves coordinating emergency services, resources, and personnel to effectively respond to natural disasters, such as earthquakes, hurricanes, floods, and wildfires. Optimizing disaster response planning is critical for efficiently allocating resources, minimizing response times, and saving lives. However, disaster response planning is a complex and dynamic problem, with uncertainties in disaster severity, location, and timing.

8.4.2.2 How MFO Can Help

MFO's ability to strike a balance between exploration and exploitation makes it well-suited for the dynamic and uncertain nature of disaster response planning. The algorithm can explore different resource allocation strategies and evacuation plans while exploiting successful solutions to optimize key disaster response performance metrics.

8.4.3 Application of MFO in Supply Chain Optimization

8.4.3.1 Problem Statement

Supply chain management involves the coordination of various activities, resources, and stakeholders to ensure the smooth flow of goods and services from suppliers to end consumers. Optimizing supply chain operations is essential for minimizing costs, reducing lead times, improving customer satisfaction, and enhancing overall efficiency. However, supply chain optimization is a complex and dynamic problem due to the involvement of multiple interconnected processes and uncertainties in demand, supply, and transportation.

8.4.3.2 How MFO Can Help

MFO's ability to balance exploration and exploitation makes it suitable for tackling the multidimensional and dynamic nature of supply chain optimization. The algorithm can explore different supply chain configurations and decision-making strategies while exploiting successful solutions to optimize key supply chain performance metrics.

8.4.3.3 Future Research

Future research in this area could focus on integrating real-time data and situational awareness into the optimization model, exploring multi-objective optimization to balance conflicting disaster response objectives and considering community-level engagement in disaster planning. Additionally, exploring hybrid approaches that combine MFO with machine learning techniques for demand forecasting and resource allocation could further improve disaster response planning.

Future research in this area could focus on incorporating additional factors into the optimization model, such as sustainability considerations, supplier performance, and demand forecasting. Moreover, exploring hybrid approaches that combine MFO with machine learning techniques for demand prediction and supply chain risk management could further enhance the effectiveness of supply chain optimization.

The MFO algorithm has shown promising capabilities in various optimization scenarios, but its journey is far from over. As researchers continue to push the boundaries of metaheuristics, several intriguing avenues of future research emerge to further enhance and explore the potential of MFO. Some potential areas of future research include:

Advanced Parameter Adaptation: Investigating more sophisticated and adaptive parameter settings for MFO could lead to improved performance in handling complex and dynamic optimization landscapes. Dynamic parameter adaptation strategies could be explored to enable MFO to dynamically adjust its parameters during the optimization process, enhancing its responsiveness and efficiency.

Hybridization with Other Algorithms: Exploring hybridization with other optimization algorithms could be fruitful in creating novel and powerful optimization approaches. Combining the strengths of MFO with those of other metaheuristics, such as GAs, PSO, or ant colony optimization, could result in hybrid algorithms with enhanced global search capabilities.

Multi-Objective Optimization: Extending MFO to address multi-objective optimization problems is an exciting avenue for future research. Developing a multi-objective variant of MFO that can efficiently find a set of Pareto-optimal solutions would cater to a wide range of real-world applications with conflicting objectives.

8.5 CONCLUSION

In this chapter, we embarked on a transformative journey that not only unraveled the secrets of the MFO algorithm but also presented an improved and enhanced version of this mesmerizing nature-inspired heuristic. Through our quest, we sought to understand the core of MFO, while also infusing it with innovative modifications to elevate its performance and versatility.

Experiment	MFO Performance	Wireless Communication Impact
Experiment 1	Significantly improved convergence speed compared to traditional algorithms.	Accelerated convergence leads to reduced latency in data transmission.
Experiment 2	Robust optimization across multiple parameters, yielding optimal solutions.	Enhanced resource allocation and network configuration for improved signal quality.
Experiment 3	Effective handling of non-linear optimization problems with high-dimensional spaces.	Enables efficient modulation schemes and adaptive coding techniques for data transmission.
Experiment 4	Exploration of dynamic optimization scenarios, maintaining adaptability.	Adaptable channel allocation strategies, ensuring seamless communication amidst changing network conditions.
Experiment 5	Robustness against noise and interference, maintaining stability in optimization.	Noise-resilient communication protocols, ensuring reliable data transfer even in challenging environments.
Experiment 6	Scalability demonstrated in large-scale optimization challenges.	Facilitates efficient scaling of communication networks to accommodate growing user demands.
Experiment 7	Comparative analysis showcasing competitive performance against state-of-the-art methods.	Offers a competitive edge in optimizing communication systems for higher efficiency and throughput.

In the realm of wireless communication, the MFO algorithm emerges as a formidable contender, exhibiting remarkable performance and transformative impact across a spectrum of experiments. By harnessing MFO's innate ability to navigate complex optimization landscapes, wireless communication systems witness a paradigm shift characterized by accelerated convergence, enhanced resource allocation, and adaptability to dynamic network conditions. The fusion of MFO's prowess with wireless communication engenders a synergy that propels optimization frontiers, ensuring seamless data transmission, improved signal quality, and noise-resilient communication protocols. As the experiments illuminate, MFO not only excels in refining traditional optimization challenges but also serves as a potent catalyst for revolutionizing wireless communication, driving connectivity and efficiency to unprecedented heights.

The origin of MFO lies in the enchanting behavior of moths drawn to luminous flames, a captivating inspiration that led to the development of this powerful optimization algorithm. As we immersed ourselves in the exploration of MFO, we recognized the brilliance of its fundamental components: the moth population symbolizing diversity and exploration, the magnetic attraction mechanism reflecting convergence toward optimal solutions, the dance of flames representing refining and optimization, and the soaring flight equations navigating the search space with grace.

Drawing from the spirit of exploration, we endeavored to enhance the MFO algorithm by incorporating dynamic parameter adaptation strategies. By allowing the algorithm to adapt its parameters dynamically during the optimization process, we sought to improve its robustness and responsiveness to different problem landscapes. The introduction of dynamic parameter settings opens new avenues for MFO to navigate complex and dynamic optimization scenarios, showcasing its adaptability and resilience.

Our research also sought to validate the improved MFO through rigorous performance evaluation. We selected a diverse set of benchmark functions and compared the results with the original MFO and other state-of-the-art metaheuristics. The empirical evidence showcased the superiority of the enhanced MFO, demonstrating its competitive performance and potential in tackling challenging optimization problems.

As we explored the landscape of practical applications, the enhanced MFO revealed its transformative power in various domains. From engineering design optimization to logistics and scheduling, and from financial portfolio management to image processing, the improved MFO showcased its versatility and efficiency in addressing real-world challenges.

Our journey through the research landscape has illuminated the potential of the enhanced MFO algorithm, an embodiment of the harmony between nature's wisdom and human ingenuity. As we conclude this chapter, we are filled with excitement for the future of optimization, with the enhanced MFO poised to inspire new frontiers of research and application.

In this age of dynamic problem-solving, the fusion of innovative modifications and nature's brilliance promises a bright future for optimization algorithms. As researchers and practitioners continue to push the boundaries of optimization, the enhanced MFO stands as a guiding light, illuminating the path to global optimization triumphs and leaving a lasting impact on the fascinating world of metaheuristics.

REFERENCES

1. Das, S., and Suganthan, P. N. (2011). Differential evolution: A survey of the state-of-the-art. *IEEE Transactions on Evolutionary Computation*, 15(1), 4–31.
2. Kennedy, J., and Eberhart, R. (1995). Particle swarm optimization. *Proceedings of the IEEE International Conference on Neural Networks*, 4, 1942–1948.
3. Dorigo, M., Maniezzo, V., and Colorni, A. (1996). The ant system: Optimization by a colony of cooperating agents. *IEEE Transactions on Systems, Man, and Cybernetics-Part B: Cybernetics*, 26(1), 29–41.
4. Luo, R., Jiao, L., and Gong, W. (2019). Firefly algorithm: Recent advances and applications. *Applied Sciences*, 9(22), 4931.
5. Liang, J. J., Qin, A. K., Suganthan, P. N., and Baskar, S. (2006). Comprehensive learning particle swarm optimizer for global optimization of multimodal functions. *IEEE Transactions on Evolutionary Computation*, 10(3), 281–295.
6. Yang, X. S. (2016). Moth flame optimization algorithm: A novel nature-inspired heuristic for global optimization. In: Yin, H., Gao, Y., and Zeng, Z. (Eds.), *Advances in Swarm Intelligence*. Springer, Cham, pp. 25–36.
7. Gandomi, A. H., Yang, X. S., and Talatahari, S. (2013). *Metaheuristic Applications in Structures and Infrastructures*. Springer, Berlin, Heidelberg.

CHAPTER 9

Optical Communication and Signal Process with AI

Sonam Gour, Priya Chaudhary, and Amit Rathi

9.1 INTRODUCTION

The main aim of optical antennas is nanoscale light control and manipulation. So, another name of antenna is nano-antenna or optical nano-antenna. They are very comparable to radiofrequency antennas even though they operate in the optical or near-infrared regions of the electromagnetic spectrum. Noble metals such as gold, silver, or aluminum are commonly used for these antennas because of their intense light interaction at the nanoscale, which is attributed to their outstanding plasmonic properties [1]. By concentrating electromagnetic fields into minuscule spaces, they facilitate optical nanoscale manipulation at the nanoscale and entrapment of particles in optical antennas for enhancing the sensitivity of sensors and boosting the performance of photodetectors.

Focusing the light into subwavelength volumes is used as a main characteristic of optical antennas. It allows light to be manipulated beyond the diffraction limit. This feature has enabled applications in spectroscopy, nanoscale imaging, and the development of novel optoelectronic devices [2]. Main point for the research of the optical antenna is the focus on improving their bandwidth, efficiency, and integration with nanophotonic components for varieties of applications in imaging, telecommunications, sensing, and quantum information processing.

DOI: 10.1201/9781003472506-9

9.2 DESIGN REQUIREMENTS OF OPTICAL ANTENNA

Various applications are the deciding parameters for the design of optical antennas. Some major applications are discussed in the following section:

9.2.1 Plasmonic Properties

The choice of material affects the antenna's plasmonic properties, including resonance frequency and field enhancement. Nobel metals like gold, silver, or aluminium are used for designing the optical antenna.

9.2.2 Geometry

The antenna's size, shape, and component arrangement determine a lot of its optical characteristics. Nanorods, nanodiscs, bowtie antennas, and nanoapertures are examples of common shapes; each offers special benefits for certain uses.

9.2.3 Resonance Frequency

The frequencies at which optical antennas resonate match the wavelengths of light that they are intended to interact with. To enhance the efficacy of light absorption, scattering, or coupling, the resonance frequency must be modified [3].

9.2.4 Polarisation Sensitivity

An antenna's optical response may be impacted by the polarisation of incident light. For some applications, polarisation-selective qualities might be useful, but for others, polarisation-insensitive behaviour might be required in the antenna design.

9.2.5 Efficiency

The goal of optical antennas is to effectively transform incident light into desired optical processes, like near-field enhancement, absorption, or scattering. Reducing losses from radiation damping, absorption in the metal, and other non-radiative processes is necessary to maximise efficiency.

9.2.6 Near-Field Enhancement

The main use of antenna is concentration towards electromagnetic fields with subwavelength region in the near filed region. It is essential for spectroscopy, nonlinear optics, and sensing applications.

9.2.7 Compatibility with Fabrication

The material of choice, geometry of antenna, and specifications of substrate should all work with the fabrication method. Electron beam lithography technique focused on template-assisted and beam milling. Nanofabrication is frequently used for creating nanoscale optical antennas [4].

9.2.8 Integration

It may be necessary to connect with optical antennas for optical components like photodetectors, waveguides, or emitters in order for constructing useful devices. Besides the considerations of integration, they are used for coupling efficiency, alignment accuracy, and compatibility with existing fabrication procedures.

9.2.9 Environmental Stability

A number of factors are considered, including as humidity and temperature. The material surrounding the antenna could impact the optical characteristics of the antenna. Stability in right environmental conditions is required for a gadget to function reliably.

9.2.10 Scalability

For practical applications, optical antennas required to be scalable in size, inexpensive in manufacture, and compatible for large-scale production techniques.

All objectives must be considered with a multidisciplinary approach with combining ideas such as optics, nanofabrication, materials science, and device engineering is often required.

9.3 ADVANTAGES OF OPTICAL ANTENNA

Optical antennas are advantageous in various applications because they give a number of benefits compared to the traditional optical structure and components. Some main advantages are as follows:

9.3.1 Nanoscale Light Confinement

By concentrating light into subwavelength volumes, optical antennas can significantly strengthen the surrounding electromagnetic fields. This property enhances nonlinear optical effects, photodetection efficiency, and sensor sensitivity by permitting more light-matter interactions at the nanoscale.

9.3.2 Plasmonic Resonances

Optical antennas utilise the plasmonic resonances of noble metal nanostructures, which can be tuned to match specific light wavelengths. These resonances are helpful for wavelength-selective filtering, surface-enhanced spectroscopy, and display colour tuning because they enable light to be selectively absorbed, scattered, or amplified at particular frequencies.

9.3.3 Enhanced Light Harvesting

Optical antennas have the potential to enhance light harvesting in photovoltaic and photodetection devices by directing incident light towards active regions such as semiconductor junctions or quantum dots. Among the advantages of this improvement in terms of device performance are higher photocurrent production and sensitivity.

9.3.4 Subwavelength Imaging

Optical antennas enable subwavelength imaging techniques that surpass the diffraction limit of conventional optics. The light is directed towards subwavelength nanostructures or apertures; optical antennas are used for achieving resolution that is greater than that limited by the wavelength of light themselves. It makes possible for image features and structures at the nanoscale with high resolution [5].

9.3.5 Tailored Optical Characteristics

Optical qualities of an optical antenna can be adjusted by modifying its shape, size, and material composition. It is used in a range of frequencies such as optical trapping, biosensing, and plasmonic catalysis due to their tunability. Antennas with tunable polarisation sensitivity, optical responses, and spectrum selectivity can be tailored for a range of applications,

9.3.6 Flexibility in Integration

Waveguides, photonic circuits, and imaging platforms can be utilised with optical antennas with a variety of optical components and systems. Their compatibility with current nanofabrication processes can be used seamlessly with intricate devices and systems.

9.3.7 Dynamic Control

Optical antennas can display dynamic optical responses because they are actively controlled by exposing it into external stimuli like electric fields, temperature variations, or optical excitation [6]. With applications in reconfigurable optics and optoelectronics, this dynamic tunability makes possible reconfigurable optical systems, switchable plasmonic resonances, and adaptive meta surfaces.

9.3.8 Versatility

Applications for optical antennas can be found in a wide range of industries, such as quantum sensing, technologies, spectroscopy, imaging, and telecommunications. Because of their adaptability and multipurpose nature, they are useful components for creating cutting-edge optical systems and devices that are customised to meet particular application needs.

All things considered, optical antennas' special qualities and capacities make them intriguing options for expanding the field of nanophotonic and facilitating innovations across a range of technical fields.

9.4 PHYSICAL PROPERTY OF BIOMEDICAL ANTENNA

Although optical antennas and their radio frequency (RF) and microwave counterparts are very similar, there are important distinctions in their scaling behaviour and physical characteristics. The majority of these variations result from the fact that metals are strongly correlated plasmas that are referred to as free electron gases rather than ideal conductors at optical frequencies. Galvanic drives are also not commonly used to drive optical antennas.

Through the use of an optical antenna, a transmitter or receiver can interact with free optical radiation. An elemental quantum absorber or emitter, such as an atom, ion, molecule, quantum dot, or defect centre in a solid, is the optimal form for the receiver or transmitter [7].

The interaction between the radiation field and the emitter or absorber is improved by the antenna. Consequently, it offers the possibility of managing the interaction between light and matter at the level of a single quantum system. The receiver/transmitter's transition rates and, in the event of a significant interaction, even the energy-level structure are altered by the antenna's existence as shown in Figure 9.1. Similar to how the receiver-transmitter characteristics affect the antenna properties, it becomes clear that the two need to be thought of as a linked system.

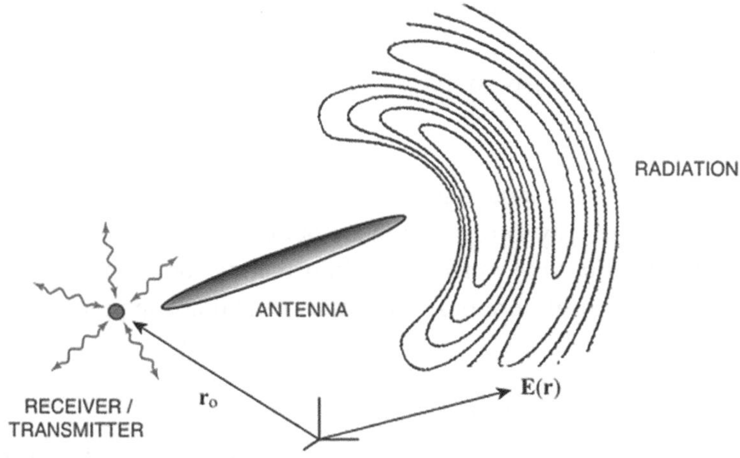

FIGURE 9.1 Optical antenna theory.

9.5 OPTICAL FIBRE-BASED ANTENNA FOR BIOMEDICAL APPLICATIONS

Because of their high sensitivity, low invasiveness, and compatibility with biological surroundings, optical fibre-based antennas have special benefits for biomedical applications. The following are some benefits and possible uses:

9.5.1 Biocompatibility

Optical fibres can come into close touch with biological tissues and fluids since they are usually composed of biocompatible materials like silica. Because of its biocompatibility, less invasive sensing and imaging methods can be used, which lessens tissue damage and patient discomfort [8].

9.5.2 Miniaturisation

The creation of minuscule antennas for targeted distribution to certain biological areas is made possible by the ability to produce optical fibres with diameters as small as a few micrometres. Miniaturisation is highly beneficial for applications that need precise spatial resolution and localisation, such as targeted drug delivery and in vivo imaging.

9.5.3 Telemedicine

The ability of optical fibre antennas to transmit and receive optical signals over great distances allows for the remote detection and observation of biological activity taking place far inside the body. This capability is invaluable in scenarios such as endoscopic imaging were large, traditional antennas may be obtrusive or impractical.

9.5.4 Multiplexing

Many physiological characteristics or biomarkers can be simultaneously imaged and sensed by integrating many optical fibres into a single probe or catheter. Multiplexing enhances the information content and diagnostic capabilities of optical fibre-based antennas, enabling comprehensive biomedical evaluations.

9.5.5 Extremely Perceptive

Because optical fibres are highly sensitive to variations in optical properties such as refractive index, absorption, and scattering, they are perfect for label-free biomolecule and cellular process detection [9]. Target analytes can be selectively collected by functionalising optical fibre antennas with specific ligands or receptors, enabling the sensitive detection of infections and disease biomarkers.

9.5.6 Live Observation

Optical fibre antennas give continuous monitoring of patient's state and their progression for illness by giving real-time input for physiological signs. Real-time monitoring is particularly helpful when it comes to dynamic processes where rapid response is necessary for patient care, such as medication metabolism, tissue perfusion, and cerebral activity.

9.5.7 Imaging at High Resolution

Utilising optical fibre antennas in conjunction with state-of-the-art imaging modalities such as multiphoton microscopy, optical coherence tomography (OCT), and confocal microscopy allows for high-resolution imaging of biological tissues and structures. The capacity of high-resolution imaging to offer accurate investigation of cellular shape, tissue architecture, and disease pathology supports early diagnosis and therapy planning. Besides being utilised for sensing and imaging, optical fibre antennas can also be employed for therapeutic applications like photodynamic therapy and phototherapy.

Optical fibre antennas have the potential to be useful in the treatment of cancer and the management of infections because they can selectively ablate tumours, eradicate infections, and activate biological processes by directing light energy directly to target areas. All things considered, optical fibre-based antennas are quite promising for a variety of biomedical uses, such as monitoring, treatments, and diagnostics. It is anticipated that further study and development in this area will produce creative answers to problems in healthcare and enhance patient outcomes.

9.6 WORKING PRINCIPAL OF OPTICAL FIBRE-BASED ANTENNA

The transmission and reception of optical signals via the optical fibre are essential to the operation of an antenna based on optical fibre. This is how it typically operates:

9.6.1 Transmission of Optical Signals

Through total internal reflection, optical fibres function as waveguides, directing light along their length. An optical signal that is fired into one end of the fibre will travel down its core with the least amount of loss from scattering or reflection. To encode information, the signal can be modified in a number of ways, including phase modulation and intensity modulation.

9.6.2 Fibre-Optic Antenna Structure

An optical fibre coupled to a photonic or plasmonic nanostructure at the end is the usual configuration for an optical fibre-based antenna. Similar to the antenna in conventional electromagnetic systems, this nanostructure functions as the radiating element. The nanostructure can be engineered to improve the way light interacts with matter, for example, by concentrating light into subwavelength volumes or forming plasmonic resonances.

9.6.3 Light-Matter Interaction

Optical fibre lights for the antennas interact with optical characteristics of the nanostructure. This interaction leads to various effects, including scattering, enhanced light emission, or absorption, all of which depend on the design of the nanostructure. Plasmonic nanostructures can focus electromagnetic fields into nanoscale volumes by inducing surface plasmon.

9.6.4 Optical Signal Radiation

As a result of interactions between light and nanostructure, optical signals can be discharged into the environment. They generated signals, which contains modulation of the incoming light. It can be detected by external optical receivers. The efficiency and radiation pattern of the antenna depend on a number of factors, such as material composition, geometry, and excitation conditions of the nanostructure.

9.6.5 Optical Signal Receiving

The optical fibre antenna can sense incoming optical signals from the surroundings when they are operated in the receiving mode. Optical properties such as resonance shifting and intensity modulation can be changed with interactions between external optical signals and the nanostructure at the antenna's end.

9.6.6 Design Specific to Applications

Optical fibres antenna can be used for specific applications, including sensing, imaging, and communication. Antennas for sensing applications can be adjusted for high sensitivity for changes in the environment surrounding, while antennas for imaging applications might be designed to achieve high contrast and spatial resolution.

Optical fibre antennas use nanostructures' unique properties for transmitting and receiving the signals.

9.7 MATERIAL DESIGN FOR OPTICAL FIBRE ANTENNAS

The application, fabrication methods, desired optical qualities, and environmental compatibility are elements that determine the construction of optical fibre antennas. The following materials are frequently used for building optical fibre antennas:

9.7.1 Silica-Based Optical Fibres

Silica is used for the most widely used waveguides for optical signal transmission. They have minimal optical losses, good mechanical flexibility, and high transparency with a wide range of wavelengths for visible and near-infrared region. Germanium or phosphorus are commonly used for adding silica fibres in order to alter their optical characteristics, including dispersion and refractive index.

9.7.2 Noble Metals (Such as Gold and Silver)

Noble metals like gold and silver are frequently utilised for creating nanostructures at the end of optical fibres for plasmonic antennas. Strong surface plasmon resonances are shown by these metals, which can improve nanoscale light-matter interactions. Commonly used nanostructures for plasmonic optical fibre antennas are bowtie antennas, nanorods, nano discs, and gold and silver nanoparticles.

9.7.3 Dielectric Materials

Polymers, silicon, silicon nitride, and titanium dioxide (TiO_2) are examples of dielectric materials that can be used to produce nanostructures on optical fibres [10]. Reduced optical losses, optical property tunability, and compatibility with complementary metal-oxide-semiconductor (CMOS) production techniques are some of the benefits of dielectric antennas. Optical fibres can be combined with dielectric nanostructures like waveguides, photonic crystals, and dielectric resonators to create tailored optical functionalities.

9.7.4 Semiconductor Materials

Silicon (Si) and gallium arsenide (GaAs) are two semiconductor materials that can be used to provide light sources (like semiconductor lasers) and photodetectors (like photodiodes) in optical fibre antennas [11]. Optical fibres and semiconductor nanostructures can be joined directly by techniques like epitaxial growth, wafer bonding, and hybrid integration.

9.7.5 Functional Coatings

To alter the optical characteristics of optical fibres or nanostructures or improve their performance for particular uses, functional coatings or layers can be put on their surface. Biomolecular coatings for biofunctionalisation and biosensing, dielectric coatings for antireflection or passivation, and photonic crystal coatings for improved light manipulation are a few examples.

9.7.6 Hybrid Materials

By combining several materials in hybrid structures, such as semiconductor-metal hybrids or metal-dielectric composites, it is possible to produce optical qualities or functions that are not possible when using the component materials alone. Hybrid optical fibre antennas can minimise the drawbacks of each component material while utilising their capabilities.

Biocompatibility (for biomedical applications), mechanical robustness, cost-effectiveness, optical performance, and compatible with fabrication processes are some of the factors that influence the selection of materials for optical fibre antennas. To maximise the performance and design of optical fibre antennas for particular applications, multidisciplinary approaches combining materials science, nanotechnology, photonics, and device engineering are frequently used.

9.8 EQUIPMENT USED FOR FIBRE-BASED ANTENNA

The tools required for fibre-based antennas vary depending on the kind of antenna being used and the particular application. The following is a list of typical equipment used:

9.8.1 Optical Fibre

The essential part of fibre-based antennas is optical fibres. Light signals are guided to and from the antenna construction by them. The fibres can differ in terms of material composition (plastic, silica), core diameter, and kind (single-mode, multi-mode).

9.8.2 Light Source

In order to produce optical signals that are sent through the optical fibre, a light source is needed. Depending on the needs of the application, this can comprise lasers, light-emitting diodes (LEDs), or broadband light sources.

9.8.3 Photodetector

A photodetector is a device that helps the antenna transform optical signals into electrical signals. Avalanche photodiodes (APDs), photomultiplier tubes (PMTs), and photodiodes are common forms of photodetectors.

9.8.4 Splitting and Combining

Coupling optical signals into and out of the optical fibre are the functions of optical couplers and splitters, respectively. They might be active (like optical modulators) or passive (like fibre couplers and splitters).

9.8.5 Optical Amplifiers

Optical amplifiers are used to increase the power of an optical signal, especially in applications requiring low-sensitivity detection or long-distance transmission. Two such examples are semiconductor optical amplifiers (SOAs) and erbium-doped fibre amplifiers (EDFAs).

9.8.6 Optical Modulators

They are used to change an optical signal's polarisation, phase, or strength as it travels through an optical fibre. These could be Mach-Zehnder interferometers (MZIs), acousto-optic modulators, or electro-optic modulators.

9.8.7 Optical Filters

Optical filters are used to filter optical signals according to particular wavelengths or frequency bands. Thin-film interference filters, fibre Bragg gratings, or Fabry-Perot cavities can be used to implement them.

9.8.8 Spectrum Analysers

They are used for measuring the bandwidth, wavelength, and power level of optical signals, among other spectral characteristics. They are crucial for describing the functionality of optical systems and fibre-based antennas. Specifically made spectrum analysers for the analysis of optical signals are known as optical spectrum analysers or OSAs for short [12]. They can identify features with tiny linewidths and offer high-resolution measurements of optical spectra.

9.8.9 Optical Microscopes

The nanostructures that are created on or around the optical fibre antenna are inspected and characterised using optical microscopes. They are necessary for alignment, quality assurance, and antenna performance optimisation [13].

9.8.10 Signal Generators and Analysers

These devices are used to produce and examine electrical signals that communicate with the antenna system's optical components [14]. They are necessary for debugging and verifying the overall functionality of the system.

9.8.11 Alignment Tools

To precisely align optical components and nanostructures within the fibre-based antenna system, alignment tools like piezoelectric stages, micromanipulators, and automated alignment systems are utilised.

These are simply a few instances of the gear that is frequently used with fibre-based antennas. Depending on the use, complexity, and performance requirements of the antenna system, different equipment will be needed.

9.9 ANTENNA FOR OPTICAL COMMUNICATION WITH SIGNAL PROCESSING

The digital signal processing deals with the modification of the optical signals after being translated from analogue to digital signal. For optical communication systems to effectively deal with the complications of light-based data transmission, signal processing is necessary. In order to improve signal quality, diminish the impact of channel difficulties, and lower noise, it includes different techniques which comprise modulation, coding-decoding, filtering, and equalisation. Signal processing provides the high data rate communication by utilising the current carrier frequencies. To improve data transfer, error correction, and overall system performance in high-speed optical communication, the signal processing handles the electric signals produced from the optical signal. Various technologies are emerged in the competition to offer ultra-fast wireless communication systems, like optical wireless communication (OWC) systems. For OWC, digital signal processing is essential to optimise data transmission efficiency and dependability in difficult circumstances, such as space, air, or underwater.

9.9.1 Optical Wireless Communication (OWC)

Optical wireless communication offers high-speed wireless data transmission across short to medium distances by using light. For data transmission across the atmosphere, OWC systems utilise an OW transmitter to modulate a beam of visible, infrared or ultra-violet light. OWC system provides a substitute for conventional RF communication and delivers solutions to radio system bandwidth constraints [15].

The near-infrared light (mainly 850, 1,310, and 1,550 nm) based antennas are very useful for data transmission in near-infrared wireless optical communication system. The OWC system transmits light using laser diodes and LEDs, comparable to fibre optic systems. Since the OWC system can work at high data rates and have a short linewidth and great efficiency, laser diodes are frequently utilised as the light source in such systems. For the laser diode as light source in antenna for near-infrared OWC, the choice of III–V semiconductor material serves as crucial, specially adapted for fibre optic communication, near-infrared spectroscopy, and biomedical applications. A few semiconductor materials such as silicon, InP, GaAs, InGaAs, GaAsSb, InPSb, and AlGaAs, are well suited for these kinds of applications and are widely used. The OWC system includes the infrared and visible communication potential approaches. The OWC system has several major types such as free space optics (FSO), underwater wireless optical communication (UWOC), and inter-satellite optical wireless communication (IsOWC), each of which offers unique benefits and opportunities.

9.9.1.1 Free Space Optics (FSO)

The free space optic transmits data via light (typically through laser) between two points with a clear line of sight. It is utilised in metropolitan areas and between buildings, where establishing a physical connection would be impractical. The FSO system is perfect for broadband access because it can travel long distances at high data rates.

9.9.1.2 Underwater Wireless Optical Communication (UWOC)

The UWOC system transmits data across water using visible light, frequently in the blue or green wavelength. Due to high attenuation or limited bandwidth, conventional techniques like radio waves are inefficient in underwater circumstances. In these situations, this technology is especially helpful. Environment monitoring, marine communication, and underwater exploration all depend on UWOC.

9.9.1.3 Inter-Satellite Optical Wireless Communication (IsOWC)

In inter-satellite OWC, laser beams are used to facilitate satellite-to-satellite communication. Deep space communication, worldwide internet coverage, and satellite networks all depend on this technology because of its exceptionally high data rates and minimal latency. In IsOWC systems, optical antennas need to be well aligned and able to resist the adverse conditions of space.

9.9.2 Role of AI in Antennas for Optical Wireless Communication

Artificial intelligence (AI) technologies like deep learning, machine learning, and neural networks can be applied to the OWC system to optimise system design, resulting in improved performance, enhanced security, and real-time adaptation to changing situations.

9.9.2.1 Optimisation of Antenna Design

AI plays a crucial role in developing high-performance antennas with appropriate features for certain OWC applications. In order to optimise antenna designs, AI systems, in particular machine learning models, can examine large datasets of material qualities, design parameters, and ambient factors. Antennas with increased gain, directivity, and efficiency result from this.

9.9.2.2 Real-Time System Adaptation

AI maintains excellent communication capabilities in a variety of environmental circumstances. Real-time analysis of the communication environment by AI can allow for the modification of the system parameters in real time, including power levels, modulation schemes, and beam steering. In spite of changing situations, this implies that the optical wireless system always runs at maximum efficiency.

9.9.2.3 Adaptive Signal Processing

AI enhances the quality and reliability of data transfer in instances of difficulty. AI has the potential to improve digital signal processing through consistent adaptation to altering the communication channel conditions. AI-driven adaptive filtering ensures decreasing error rates and improved signal reception by reducing the impact of atmospheric disturbance in FSO systems and scattering in UWOC systems.

9.9.2.4 Enhanced Security

AI ensures secure data transmission. Traffic pattern can be monitored by AI, which can identify irregularities that could point to security risks like signal jamming or spying. AI has the ability to improve encryption methods, increasing communication security against online threats.

9.10 LITERATURE REVIEW ON THE USE OF OPTICAL FIBRE ANTENNAS IN BIOMEDICAL

According to a review of the literature, there is an increasing amount of research being done on the possible uses of optical fibre antennas in biomedicine [16]. A fibre-optic Fabry-Perot interferometric sensor for very sensitive and accurate physiological parameter measurement, including blood pressure and temperature, was demonstrated in a study that emphasised the benefits of optical fibre sensors' miniaturisation, biocompatibility, and multiplexing capacities for usage in biomedical applications [17]. In order to facilitate non-invasive illness diagnostics, the author created a miniature fibre-optic confocal microscope for in vivo imaging of biological tissues with subcellular resolution.

The potential of OCT for the early identification of ocular disorders was demonstrated which developed an OCT system based on fibre-optic antennas for high-resolution imaging of retinal structures and vasculature [18]. The use of optical fibre-based antennas for controlled release of therapeutic agents in localised drug delivery applications was the subject of a study by the author which demonstrated superior temporal and spatial control over drug release compared to traditional approaches.

The effectiveness of photothermal therapy for the treatment of cancer was demonstrated by the author for investigation into the use of gold-coated optical fibres as photothermal actuators for the targeted annihilation of cancer cells.

In order to detect biomolecules like DNA and proteins label-free, the authors created a fibre-optic biosensor based on plasmonic antennas that has high sensitivity and specificity for use in biomedical diagnostics [19]. In their overview of the latest developments in fibre-optic surface-enhanced Raman scattering (SERS) sensors for biomedical sensing applications, they emphasised the sensors' promise for sensitive and quick identification of disease biomarkers.

Optogenetic manipulation of neuronal activity in deep brain areas can be achieved with fibre-optic antennas, opening up new possibilities for neurotherapeutic applications and providing precise control over neural circuits [20].

These studies highlight the increasing importance of optical antennas in a wide range of applications. The unique properties of antennas, including high sensitivity, small size, and the ability to function in biological environments, make them promising for advancing biomedical diagnosis, treatment, and research. Further research in this field is used with significant therapeutic applications.

9.11 CONCLUSION

Optical fibre antennas have a wide range of biological applications due to their unique capacities and properties. Optical fibre antennas have been useful in a variety of biomedical domains, drug administration, biological analysis, neural interface, sensing, and imaging. It is used for developing tiny, biocompatible devices which communicate with biological systems. They are perfect for integration with biomedical equipment and minimally invasive procedures because of adaptability and their small size. Optical fibre antennas are a sensitive and selective means of detecting cells, biomolecules, and tissues.

They enable label-free sensing with good detection limits and specificity and also use for enhancing light-matter interactions at the nanoscale. Fibre-based antennas are used in high-resolution imaging of biological tissues and structures. It is used for aiding in the early detection and diagnosis of disease. It offers fine-grained control by therapeutic effects at certain body sites over medicine delivery. It includes optogenetics, photothermal therapy, and targeted drug release, which may be used in regenerative medicine, neurotherapeutics, and cancer treatment. Real-time monitoring of physiological data and neural activity is made possible by fibre-based antennas.

Optical fibre antennas enable bidirectional communication with neural tissues. It is essential for the development of periprosthetic devices and brain machine interfaces. Integrating optical fibre antennas with other optical and electrical components are used for creating multifunctional biomedical devices with enhanced functionality and performance. Their compatibility with current fabrication techniques facilitates their easy integration into biological systems and technologies. The use of optical fibre antennas has greatly advanced biomedical research, diagnosis, and therapy. It is expected that additional research and development in that field enhances patient monitoring outcomes and expands our knowledge of complex biological systems.

REFERENCES

1. H. Gersen, M. F. García-Parajó, L. Novotny, J. A. Veerman, L. Kuipers, and N. F. van Hulst, "Influencing the angular emission of a single molecule," *Phys. Rev. Lett.* 85, 5312 (2000).
2. C. Huang, A. Bouhelier, G. C. des Francs, A. Bruyant, A. Guenot, E. Finot, J.-C. Weeber, and A. Dereux, "Gain, detuning, and radiation patterns of nanoparticle optical antennas," *Phys. Rev. B* 78, 155407 (2008).
3. R. Carminati, M. Nieto-Vesperinas, and J.-J. Greffet, "Reciprocity of evanescent electromagnetic waves," *J. Opt. Soc. Am. A* 15, 706–712 (1998).

4. T. H. Taminiau, F. D. Stefani, and N. F. van Hulst, "Enhanced directional excitation and emission of single emitters by a nano-optical Yagi-Uda antenna," *Opt. Express* 16, 10858–10866 (2008).
5. T. H. Taminiau and N. F. Van Hulst, ICFO—The Institute of Photonic Sciences, Castelldefels (Barcelona), Spain (personal communication, 2008).
6. K. T. Shimizu, W. K. Woo, B. R. Fisher, H. J. Eisler, and M. G. Bawendi, "Surface-enhanced emission from single semiconductor nanocrystals," *Phys. Rev. Lett.* 89, 117401 (2002).
7. J. N. Farahani, D. W. Pohl, H.-J. Eisler, and B. Hecht, "Single quantum dot coupled to a scanning optical antenna: a tunable superemitter," *Phys. Rev. Lett.* 95, 017402 (2005).
8. P. Bharadwaj and L. Novotny, "Spectral dependence of single molecule fluorescence enhancement," *Opt. Express* 15, 14266–14274 (2007).
9. R. Loudon, *The Quantum Theory of Light*, 2nd ed. (Oxford Science Publications, Clarendon, 1983).
10. L. Tang, S. E. Kocabas, S. Latif, A. K. Okyay, D.-S. Ly-Gagnon, K. C. Saraswat, and D. A. B. Miller, "Nanometre-scale germanium photodetector enhanced by a near-infrared dipole antenna," *Nat. Photonics* 2, 226–229 (2008).
11. D. Tosi, S. Poeggal, I Iordachita, and E. Schena "Fiber optic sensors for biomedical applications," in H. Alemohammad ed. *Opto-Mechanical Fiber Optic Sensors*, pp. 301–303, (Butterworth-Heinemann, 2018).
12. H. Lamela, D. Gallego, R. Gutierrez, and A. Oraevsky "Interferometric fiber optic sensors for biomedical applications of optoacoustic imaging," *J Biophotonics* 4(3), 184–192 (2011).
13. L. Chin, and W. M. Whelan "Optical fiber sensors for biomedical applications," in A. J. Welch, M. J. C. Gemert eds. *Optical-Thermal Response of Laser-Irradiated Tissue*, pp. 661–712, (Springer Science+Business Media B.V., 2011).
14. D. P. Fromm, A. Sundaramurthy, A. Kinkhabwala, P. J. Schuck, G. S. Kino, and W. E. Moerner, "Exploring the chemical enhancement for surface enhanced Raman scattering with Au bowtie nanoantennas," *J. Chem. Phys.* 124, 061101 (2006).

15. O. Alsulami, A. T. Hussein, M. T. Alresheedi, and J. M. H. Elmirghani "Optical wireless communication systems, a survey," *arXiv preprint arXiv:1812.11544* (2018).
16. R. M. Stöckle, Y. D. Suh, V. Deckert, and R. Zenobi, "Nanoscale chemical analysis by tip-enhanced Raman spectroscopy," *Chem. Phys. Lett.* 318, 131–136 (2000).
17. Hartschuh, E. Sanchez, X. Xie, and L. Novotny, "High-resolution nearfield Raman microscopy of single-walled carbon nanotubes," *Phys. Rev. Lett.* 90, 095503 (2003).
18. H. Hartschuh, H. Qian, A. J. Meixner, N. Anderson, and L. Novotny, "Tipenhanced optical spectroscopy of single-walled carbon nanotubes," in S. Kawata, V. M. Shalaev eds. *Tip Enhancement: Advances in NanoOptics and Nano-Photonics*, pp. 157–176 (Elsevier, 2007).
19. H. R. Stuart and D. G. Hall, "Absorption enhancement in silicon-on insulator waveguides using metal island films," *Appl. Phys. Lett.* 69, 2327–2329 (1996).
20. K. R. Catchpole and S. Pillai, "Absorption enhancement due to scattering by dipoles into silicon waveguides," *J. Appl. Phys.* 100, 044504 (2006).

CHAPTER 10

Microwave Photonic Radar

Introduction and Signal Processing Challenges

Hum Nath Parajuli

Radio detection and ranging (radar) systems have traditionally been used to determine the range, velocity, and imaging of target objects. Conventional radar systems rely on electronic technologies. However, recent advancements in photonic technology have generated significant interest in microwave photonics (MWP)-based radar systems. These systems provide significant benefits compared to conventional electronic radar, such as high-bandwidth and high-frequency signal generation and processing with minimal noise. Additionally, they allow reconfigurability in terms of bandwidth, chirp rate, and frequency for radar signals. This potential for advancement could transform the future of radar technology. This chapter provides an overview of MWP-based radars, emphasizing their benefits and potential for future radar systems.

Various radar signal processing techniques are essential for accurately extracting desired measurements. These techniques include waveform design tailored to specific applications; mitigation of path loss, noise, and interference; and enhancements in range, Doppler, and imaging performance. Discussing all of these signal-processing challenges is beyond the scope of this chapter. However, this chapter introduces radar-radar interference mitigation (IM) techniques for photonic radar systems, which

DOI: 10.1201/9781003472506-10

are increasingly important as the number of radar users are growing with advancements in technology in diverse applications. This chapter addresses the interference issue by exploring two key strategies: designing interference-resistant waveforms and applying deep learning techniques to effectively suppress the effects of multi-radar interferences. These techniques hold significant potential for advancing radar technology and addressing the emerging challenges of multi-radar IM.

10.1 INTRODUCTION TO RADAR SYSTEMS

Radio detection and ranging (radar) sensing systems use electromagnetic waves across a wide frequency range, from long-range (10 MHz to 1 GHz) to short-range (1 GHz to mmWave), to determine the range, velocity, direction, and imaging of target objects. Depending on the operating frequency bands, which range from the HF-band (3–30 MHz) to mm-wave bands (40–300 GHz), radar applications vary [1]. The initial application of radar dates back to World War II, where it was used for military purposes to locate target objects. Since then, radar systems have been utilized in various fields, including defense, remote sensing, industry, air navigation, marine, and more recently, automotive, agriculture, and medical industries [2–7].

Radars can be categorized based on operating frequency, waveform type, and detection methods. For instance, pulse radar uses short pulses to detect target objects and is employed in military, airport, and weather monitoring applications. Continuous wave (CW) radar uses a continuous signal to measure the frequency shift of the echo signal (reflected signal from an object), which helps to determine the range and velocity of the target object. A subtype of CW radar is a frequency-modulated continuous wave (FMCW) radar, which uses a continuous waveform with linear frequency modulation (LFM) [8–10]. LFM radar is popular in automotive applications due to its excellent pulse compression characteristics, providing finer range and Doppler (velocity) resolution. Similarly, synthetic aperture radar (SAR) uses high-frequency CW signals along with antenna movement to capture high-resolution images of target objects, making it popular in remote sensing applications.

In a typical radar system, the radar transmitter generates a waveform or pulse that is directed toward target objects, and the reflected echo is captured by the radar receiver. By analyzing the properties of the received echo signal, such as amplitude, phase, frequency, and time delay, various measurable characteristics of the detected object, including range, velocity, direction, shape, and size can be determined. Various radar signal processing techniques are essential for accurately extracting these desired properties. These techniques include designing suitable waveforms tailored to specific applications; minimizing the effects of path loss, noise, and interference; and enhancing range, Doppler, and imaging measurement performance through advanced signal processing methods, including deep learning (DL) techniques.

Conventional commercial radar systems primarily rely on electronic technology. However, with recent advancements in photonic technology, there has been significant interest in developing MWP-based radar systems, especially within academic research. Due to the numerous advantages photonics offers over electronic systems, it holds substantial potential to revolutionize future radar systems. The next section provides a brief introduction to MWP-based radars.

10.2 MICROWAVE PHOTONICS (MWP) FOR RADAR

MWP in radar involves the integration of photonic and microwave technologies to generate, transmit, and process microwave radar signals, typically within the frequency range of 1–100 GHz. This interdisciplinary approach leverages the advantages of photonics, such as low-loss transmission, wide bandwidth, and immunity to electromagnetic interference, to enhance radar system performance. MWP-based radar systems can operate in free space optical (FSO) channels, where they benefit from the same advantages but also suffer from similar environmental challenges as traditional FSO systems [11–13]. These challenges include atmospheric turbulence, absorption, scattering, and weather conditions such as fog, rain, and snow, all of which can impact signal quality and overall system reliability [14–19]. Despite these challenges, MWP techniques are increasingly important in modern radar systems, offering improved signal quality, greater flexibility in system design, and the potential for miniaturization and integration with other photonic systems.

For radar applications, MWP techniques offer reconfigurable high-bandwidth and high-frequency signal generation, down-conversion, up-conversion, de-chirping, and beam-forming functionalities [20–22].

Among these benefits, the high-bandwidth signal generation is particularly important, as range resolution is a measure of the ability to distinguish between objects and depends on the radar signal's bandwidth, given by $\Delta R = \frac{c}{2B}$, where c is the speed of light and B is the bandwidth of the radar signal. The higher the bandwidth of the radar signal, the finer the range resolution, allowing for more precise measurements [23–26]. In contrast, achieving these capabilities with electronics is challenging, as electronics suffer from high path loss, electromagnetic interference, increased noise at higher frequencies and high-bandwidth signals, and limited reconfigurability for high-frequency operations.

10.2.1 MWP for Reconfigurable Radar Signal Generation

MWP offers the generation of arbitrary radar waveforms with reconfigurable parameters such as frequency, bandwidth, phase, and chirp rate. In the photonic domain, a radar signal with a bandwidth of 30 GHz has already been demonstrated [27]. High-bandwidth signals are crucial for achieving finer range resolution in object detection, especially when using LFM waveforms. The common techniques for photonics-based radar signal generation can be given as:

10.2.1.1 Photonic Heterodyning

This technique creates an optical signal with a beat frequency that can be transformed into a microwave signal by utilizing two lasers with different wavelengths or one mode-locked laser with multiple wavelengths. Various radar waveforms, including LFM, can be produced by varying and beating the wavelengths at the photodetector [28–31]. High stability, minimal phase noise, and a variety of waveforms can be achieved using the photonic heterodyning technique [32].

10.2.1.2 Photonic Multiplication

In such techniques, electro-optic modulator-generated sidebands are used to generate radar waveforms. Intended radar waveform with low frequency and bandwidth is fed to the modulator, which produces higher-order sidebands. Upon the beating of the higher-order sidebands in the photodetector, the microwave radar signal with high frequency and bandwidth can be generated. These techniques offer advantages such as high efficiency, low noise, and the ability to generate complex radar waveforms with high frequency and bandwidth multiplication. These techniques have the potential to enhance the performance and capabilities of radar systems for a variety of applications [33,34].

10.2.1.3 Optical Injection Technique

In this microwave radar signal generation technique, two laser beams beat in such a way that the injection of one beam causes a shift in the emission wavelength of the other lase. Upon injecting the external optical beam, also called master laser into a semiconductor laser, also called slave laser with varying intensity and wavelength, the redshift phenomenon occurs and the microwave signal can be generated upon beating the injected wavelength with the red-shifted wavelength of the semiconductor laser. This phenomenon can be used to generate radar waveforms with reconfigurable frequencies, bandwidths, and chirp rates [35–37].

10.3 SIGNAL PROCESSING IN PHOTONIC RADAR SYSTEM

In photonics-based radar (Ph-radar), the electronic signal processing tasks are similar to those in electronics-based radar. These tasks include designing radar waveforms tailored to specific applications, de-noising received echo signals, de-chirping, adaptive filtering, enhancing performance parameters such as range-Doppler and imaging resolution, and mitigating the effects of external radar interference. Discussing all of these signal processing challenges is out of the scope of this chapter. However, this chapter introduces radar-radar IM techniques for photonic radar systems, which are increasingly important due to the growing use of radars both for present time and in the future. These techniques are expected to advance radar technology and address emerging challenges.

10.3.1 Introduction to Radar-Radar Interference

Today, radars are widely used in a variety of fields, including remote sensing, defense, agriculture, industry, automotive applications, and consumer electronics. As radar applications increase, signals from external radar sources such as interference radar (IR) can disrupt legitimate radars, known as victim radars (VR). The serious issue arises when IR operates at the same or nearby frequency bands as the VR. This IR signal can severely affect the VR's performance by adding noise and creating false targets. In a multi-radar environment, such interference can degrade the VR's detection capabilities and pose risks to the safe and effective operation of radar systems. Therefore, mitigating interference in a multi-radar environment is essential.

Among the various radar waveforms, the LFM is the most prevalent. It can be described by:

$$T_x(t) = A_{tx} \cos\left(\left(2\pi\left(f_c - \frac{B}{2}\right)t + \pi\alpha_{tx}t^2\right)\right), \text{ for } 0 \le t \le T \quad (10.1)$$

where A_{tx} represents the amplitude of the transmitted signal, f_c is the center frequency, B denotes the operating bandwidth, α_{tx} is the chirp rate (B/T), and T is the time duration of the LFM signal. Over the period $0 \le t \le T$, the frequency of the transmitted signal increases linearly from $f_c - \frac{B}{2}$ to $f_c + \frac{B}{2}$. When this transmitted signal reflects off K number of stationary targets, the received signal $R_x(t)$ can be expressed as

$$R_x(t) = \sum_{k=1}^{K} A_{Rk} \cos\left(\left(2\pi\left(f_c - \frac{B}{2}\right)(t - \tau_{dk}) + \pi\alpha_{tx}(t - \tau_{dk})^2\right)\right), \quad (10.2)$$

where A_{Rk} is the amplitude of the reflected signal from the kth target, and τ_{dk} represents the time delay due to the distance between the radar and the kth target. The range resolution of the LFM signal, which indicates the ability to distinguish between two target objects, is proportional to the frequency shift and can be given as $\Delta R = \frac{c}{2B}$, where c is the speed of light and B is the bandwidth of the radar signal.

Now, consider the IR signal, which is also LFM and acts on the receiver. It can be expressed as,

$$R_I(t) = \sum_{y=1}^{Y} A_{Iy} \cos\left(\left(2\pi\left(f_c - \frac{B_{Iy}}{2}\right)(t - \tau_{Iy}) + \pi\alpha_{Iy}(t - \tau_{Iy})^2\right)\right) \quad (10.3)$$

where A_{Iy} is the amplitude of the yth IR signal, α_{Iy} is the chirp rate, and τ_{Iy} is the delay of the IR signal relative to the VR signal. The chirp rate

(α_{Iy}) of the IR signal is calculated as $\frac{B_{Iy}}{T_{Iy}}$, with B_{Iy} being the bandwidth and T_{Iy} the time duration of *y*th IR signal. When the chirp rates of the VR signal and IR signal are identical ($\alpha_{Iy} = \alpha_{tx}$), this is termed coherent interference; if they differ ($\alpha_{Iy} \neq \alpha_{tx}$), it is called non-coherent interference. With the presence of IR signals, whether coherent or non-coherent, receiver's output is a mix of the VR's echo and the IR signal. For simplicity, consider a scenario with one VR and one IR source. The beating frequency caused by the IR signal $R_I(t)$, after de-chirping, is given by,

$$f_{bi} = \left(\alpha_{Iy} - \alpha_{tx}\right)t + \alpha_{Iy}\tau_{Iy} \tag{10.4}$$

Figure 10.1 depicts the LFM VR signal and its de-chirped signal at the receiver in the presence of an IR signal with either the same or different chirp rates. In Figure 10.1a-i, where the IR signal has the same chirp rate as the VR signal, two constant beat frequencies f_{bo} and f_{bi} are observed at the receiver's output as shown in Figure 10.1a-ii. These beat frequencies fall within the bandwidth of the receiver and indicate the presence of the two target objects, including one ghost target as illustrated in the range profile of Figure 10.1a-iii. In Figure 10.1b-i, where the chirp rates are different, a constant beat frequency, f_{bo}, from the actual target object is observed alongside a varying beat frequency $f_{bi}(t)$ as shown in Figure 10.1b-ii. This varying beat frequency introduces additional undesired frequency components in the received signal and increases the noise floor in the range profile as depicted in Figure 10.1b-iii. Both types of interference, whether coherent, Figure 10.1a or non-coherent, Figure 10.1b, significantly impair the VR's detection performance in a multi-radar environment by creating ghost targets or increasing the noise floor. Figure 10.1 demonstrates a scenario with one target object and one interference signal acting in the received echo. But, in real-world applications, multiple object detection should be carried out along with the various interference signals. Various approaches can be implemented to minimize the effect of IR signals in a VR signal while operating in a multiple IR sources environment. This chapter introduces the techniques of mitigating IR effect through the design of interference-resistant radar waveform and through DL-based solutions.

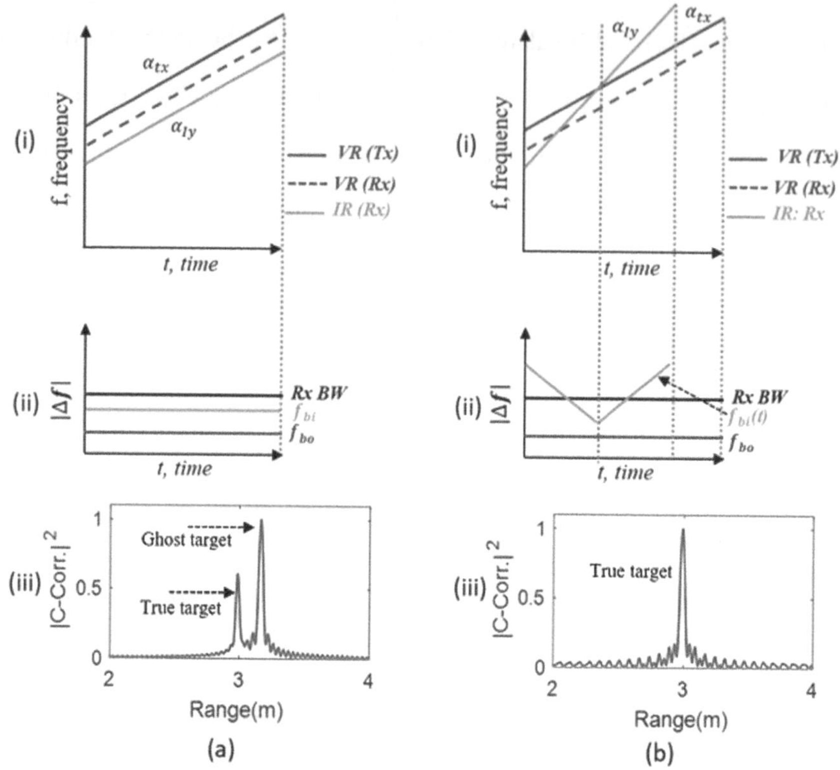

FIGURE 10.1 Depictions of interference effects: (a) coherent interference, (b) non-coherent interference. (i) F-t diagrams of echo plus interference signals, (ii) f-t diagrams of generated beat signals within the receiver, and (iii) resulting range profiles [38].

10.3.2 Radar-Radar Interference Mitigation

Among various radar waveforms, the LFM waveform is popular for its superior pulse compression abilities, which enhance range-Doppler resolution. Nevertheless, the LFM waveform shows poor interference resistance, particularly when an IR signal interferes coherently with a VR signal. Most current IM approaches depend on sophisticated signal processing methods within the radar's receiver. Consequently, designing radar waveforms with inherent IM capabilities is crucial for real-time objects detection. This chapter introduces the concept of interference-resistant waveform design. Additionally, recent advancement in DL-based IM techniques have gained attention to the radar research community, as these techniques can reconstruct the echo signal by suppressing the interference regardless of the types of VR and IR signal involved. Therefore, in this chapter, a DL-based IM technique for photonics-based radar systems is introduced.

10.3.2.1 Interference-Resistant Waveform Design

In a multi-radar environment, interference is a particularly significant factor contributing to performance degradation. As radars become more widely used across various applications, the operation of multiple radars with overlapping frequency ranges leads to interference. Interference increases the noise floor and introduces the ghost target objects into the range profile, which results in reduced detection capabilities and erroneous detections. To mitigate interference, various computationally complex signal processing techniques have been developed [31,39–44]. Though these methods can lead to latency and increased costs, interference-resistant transmission waveforms offer effective solutions to these problems. In electronics, examples of such solutions include random noise waveforms [45], phase-coded waveforms [46], multiple-chirped waveforms [47], and more. In the photonics domain, techniques like sparse-stepped frequency chirps [48] and chaotic radar [49] solutions have been proposed. Additionally, range-Doppler resolution is also a critical metric for detection accuracy. LFM waveform is popular due to its excellent compression characteristics which provide superior range-Doppler resolution. However, LFM waveform is less effective at resisting interference, particularly when the IR operates coherently to the VR. Thus, modified LFM waveforms are essential for achieving both excellent range-Doppler resolution and effective IM.

Parajuli et al. [50] introduced a modified LFM waveform known as photonically generated orthogonal up-down-chirp pair-based multi-chirp LFM (Ph-O-MLFM) waveform as an interference-resistant waveform while ensuring excellent resolution capability. This waveform is distinctive due to its orthogonal properties and the use of concatenated LFM waveforms with different chirp rates. The orthogonality of LFM signal enhances compression characteristics compared to using individual LFM signals [39,51]. This feature is incorporated into the Ph-O-MLFM waveform through the concatenation of up-down LFMs. Additionally, the use of concatenated LFM waveforms with varying chirp rates prevents coherent interference by shifting the signal-to-interference beating frequency outside the receiver's bandwidth. As a result, the use of multiple concatenated LFMs effectively reduces the impact of interference in the range profile.

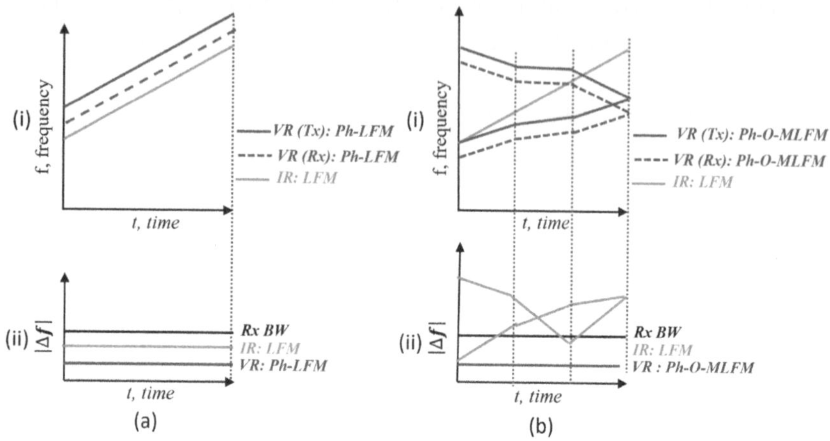

FIGURE 10.2 Diagrams showing the effects of LFM interference radar on victim radar waveforms: (a) Ph-LFM waveform and (b) Ph-O-MLFM waveform. (i) F-t diagrams of echo plus interference signals and (ii) resulting f-t diagrams of beat signals within the receiver [50].

Figure 10.2 illustrates how IR LFM affects the beat frequencies (Δf) in the radar receiver (Rx) for two different radar waveforms: common photonics-based LFM (Ph-LFM) waveform and the Ph-O-MLFM waveform. Figure 10.2a-i and b-i show the transmitted, echo, and interference signals received for the Ph-LFM and Ph-O-MLFM VR waveforms, respectively. As depicted in Figure 10.2a-ii, the Ph-LFM waveform experiences significant coherent interference because the signal-to-interference beat frequency falls within the radar Rx BW, making it difficult to suppress interference. Conversely, Figure 10.2b-ii demonstrates that for the multi-chirp Ph-O-MLFM waveform, most of the signal-to-interference beat frequency falls outside the Rx BW, thereby inherently diminishing the impact of the interference.

To demonstrate the detection capabilities of the Ph-LFM and Ph-O-MLFM waveforms, the authors set up a simulated multi-radar environment where both waveforms were generated and utilized for object detection evaluation. For the Ph-LFM, a 2-GHz, 1-μs LFM signal was used, while the Ph-O-MLFM waveform was generated using two 2-GHz, 1-μs, three-chirp LFM signals fed into a dual-drive Mach-Zehnder modulator (DDMZM). By adjusting the DDMZM parameters, photonic-based bandwidth synthesis was achieved, resulting in 4-GHz, 1-μs, Ph-LFM and Ph-O-MLFM radio frequency (RF) signals after photodetection. These waveforms are then combined with varying chirp rate 4-GHz LFM signals, and a detection analysis for two objects spaced 6 cm apart was conducted using a matched filtering technique.

Figure 10.3 illustrates the setup for generating and analyzing Ph-LFM and Ph-O-MLFM RF signals in a multi-radar environment. This setup includes an MWP system based on DDMZM to produce RF radar signals through the photonic beating of the sidebands created by the DDMZM. The setup uses a CW laser operating at 0 dBm output power with a 10 MHz linewidth, and a photodetector with a thermal noise current spectral density of $10e^{-12}$ $A/Hz^{1/2}$ to simulate realistic noise conditions. The RF signal from the photodetector is then amplified using an RF amplifier with a 20-dB gain and a noise current spectral density of $20e^{-12}$ $A/Hz^{1/2}$ and filtered through a fourth-order Bessel bandpass filter.

In the multi-radar setup depicted in the dotted box in Figure 10.3, the VR unit, IR unit, and objects space unit are configured. The VR unit transmits the photonically generated RF Ph-LFM and Ph-O-MLFM waveforms and receives echoes from two objects separated by $\Delta L = 6\,\text{cm}$. The IR unit generates 4-GHz LFM interference signals with varying chirp rates and periods, which are directed toward the VR unit's receiver (Rx). These IR signals are mixed with the echoes, and additive white Gaussian noise (AWGN) with an SNR of 0 dB simulates the reception of noisy echoes and interference signals. Object detection in the presence of LFM interference is assessed by analyzing the normalized cross-correlation (C-Corr.)

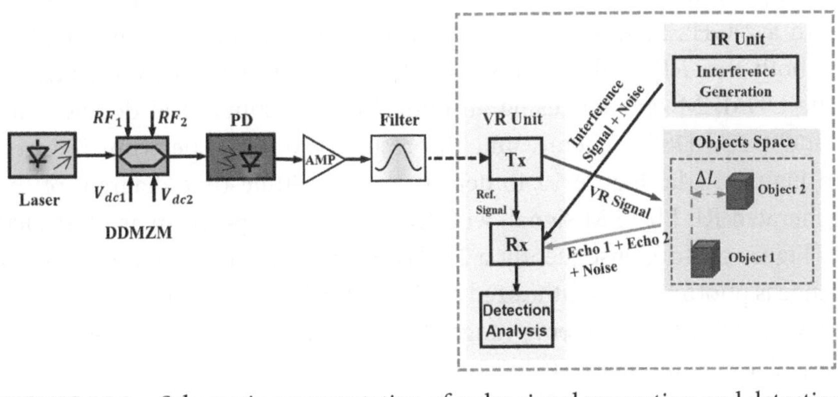

FIGURE 10.3 Schematic representation of radar signal generation and detection analysis under the influence of interfering radar signals [50].

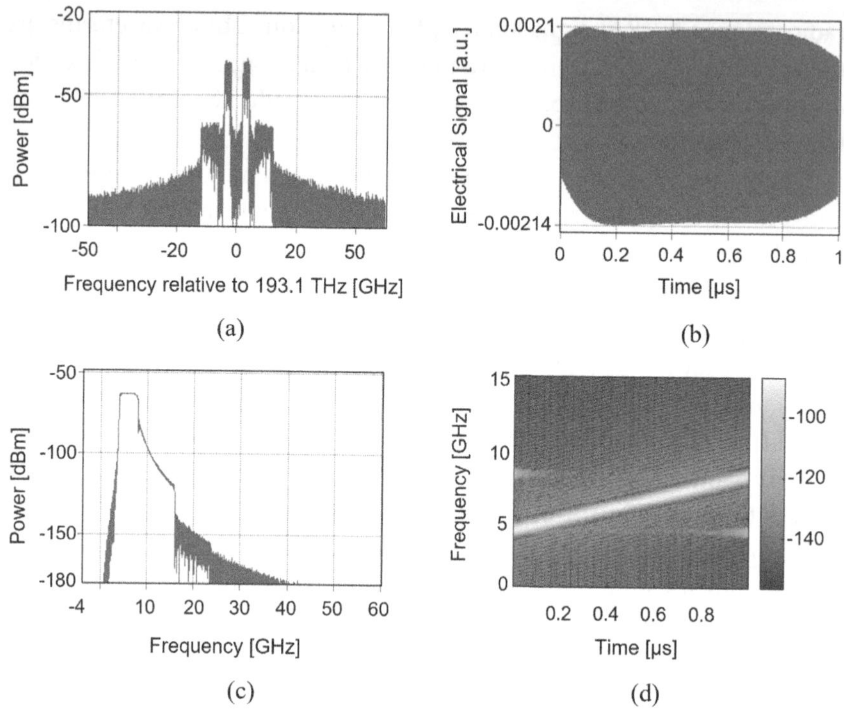

FIGURE 10.4 Ph-LFM signal generation. (a) Optical spectrum at the output of DDMZM, (b) amplitude-time diagram of the signal after photodetector, (c) RF spectrum of the generated signal after filtering, and (d) frequency-time plot of the generated signal [50].

amplitude, range resolution, and peak-to-sidelobe-level power (PSL) values in the obtained range profile.

To generate a 4-GHz Ph-LFM signal, a 2–4 GHz LFM driving signal with a 2-GHz bandwidth and a 1-µs time period is created and applied to both the RF_1 and RF_2 ports of the DDMZM with identical phases. The DDMZM operates using an optical-carrier suppressed double side-band (OCS-DSB) scheme, and the resulting optical spectrum is given in Figure 10.4a. Figure 10.4b depicts the amplitude-time diagram of the generated RF Ph-LFM signal, with the filtered RF spectrum as illustrated in Figure 10.4c. Consequently, a 2-GHz driving signal within the 2–4 GHz range is photonically synthesized by the DDMZM, generating a 4-GHz Ph-LFM signal with a frequency range of 4–8 GHz, as shown in the frequency-time diagram in Figure 10.4d.

To generate a 4-GHz Ph-O-MLFM signal, three up-chirp LFM signals with frequency ranges of 4–5.25, 5.25–5.5, and 5.5–6 GHz are concatenated and applied to the RF_1 port of DDMZM. Similarly, the RF_2 port of DDMZM is fed with three concatenated down-chirp LFM signals with frequency ranges of 8–6.75, 6.75–6.5, and 6.5–6 GHz. The up-chirp signals (4–6 GHz) and down-chirp signals (8–6 GHz) both have chirp rates of 3.75, 0.75, and 1.5 GHz/μs but with opposite signs. Each signal has a bandwidth of 2 GHz and a time period of 1 μs. In this case, the DDMZM is used in an optical double sideband (O-DSB) mode. By applying two 2 GHz driving signals to the RF1 and RF2 ports of the DDMZM, up-chirp and down-chirp signal pairs with opposite chirp rates are photonically combined, resulting in a 4-GHz, 1-μs Ph-O-MLFM RF signal after photodetection. The optical spectrum of the generated signal is shown in Figure 10.5a, and the amplitude-time diagram is presented in Figure 10.5b. Figure 10.5c depicts

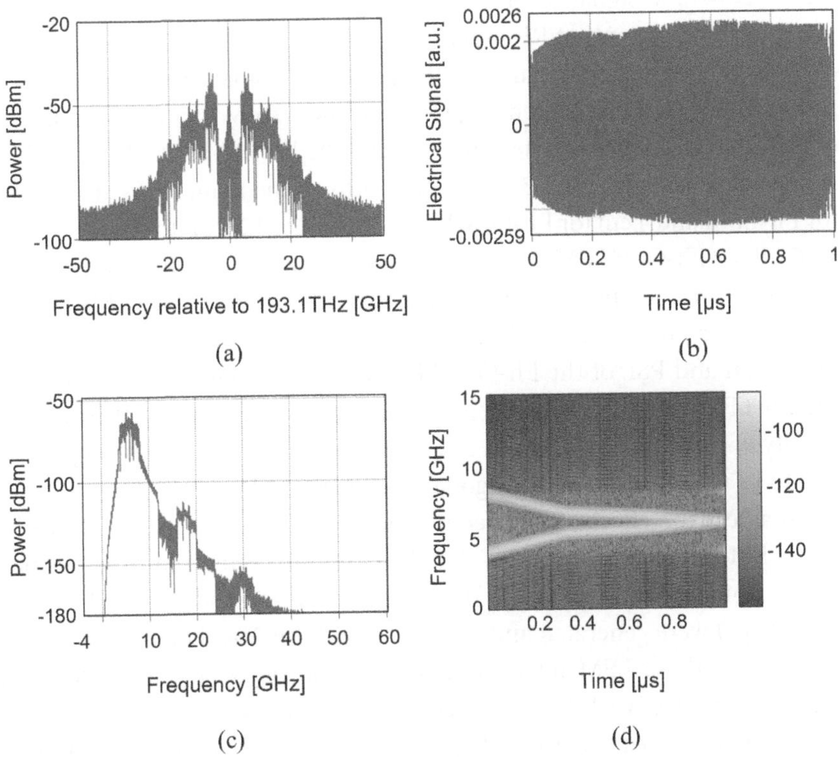

FIGURE 10.5 Ph-O-MLFM signal generation. (a) Optical spectrum at the output of DDMZM, (b) amplitude-time diagram of the signal after photodetector, (c) RF spectrum of the generated signal after filtering, and (d) f-t plot of the generated signal [50].

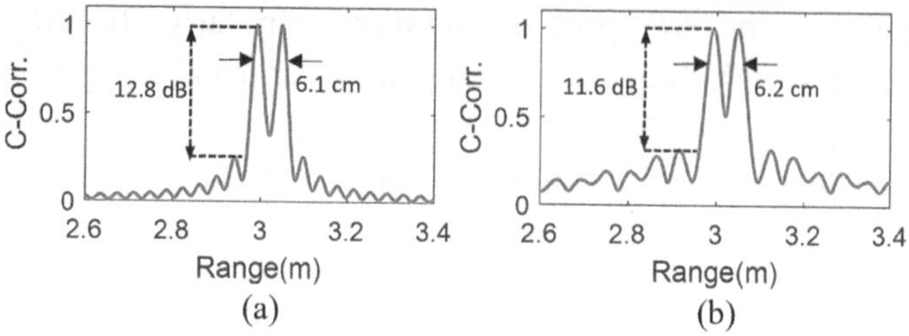

FIGURE 10.6 Range profile results for object detection without interference using (a) Ph-LFM and (b) Ph-O-MLFM signals [50].

the filtered RF spectrum, while Figure 10.5d illustrates the frequency-time diagram of the Ph-O-MLFM signal. Using the photonics-based bandwidth synthesis with DDMZM, a 4-GHz RF output signal can be produced from a 2-GHz input RF signal in the case of Ph-LFM signal generation. Similarly, for Ph-O-MLFM signal generation, two 2-GHz input RF signals are used to generate a 4-GHz output RF signal.

To evaluate the performance of the generated signals, two object detection analyses were conducted. Figure 10.6a and b show the range profiles of objects detected without interference using the Ph-LFM and Ph-O-MLFM signals, respectively. The −3 dB pulse widths of the main lobes are 3.8 cm for Ph-LFM and 4 cm for Ph-O-MLFM. The PSL is 12.8 dB for Ph-LFM and 11.6 dB for Ph-O-MLFM. Object detection accuracy is within a maximum error of 2 mm. Despite the inclusion of dissimilar chirp rate LFMs, which can enhance phase noise and degrade PSL and range resolution, the range resolution and PSL of the Ph-O-MLFM signal are comparable to those of the Ph-LFM signal. This performance is largely attributed to the orthogonality of the Ph-O-MLFM waveform, which reduces sidelobes' amplitude due to its composition of orthogonal up-down LFM pairs.

To assess the effect of interference on the generated Ph-LFM and Ph-O-MLFM VR signals, LFM-IR signals with chirp rates and time periods of I_{LFM1} (4 GHz/μs, 1 μs), I_{LFM2} (8 GHz/μs, 0.5 μs), and I_{LFM3} (16 GHz/μs, 0.25 μs) were generated and combined with the VR signals. Figure 10.7 shows how these LFM interferences affect the range profile of Ph-LFM signal. Figure 10.7a-i, b-i, and c-i depict the frequency-time diagrams of the echo signals in the presence of I_{LFM1}, $\mathrm{I}_{\mathrm{LFM2}}$, and I_{LFM3} interference signals, respectively. The range profile in Figure 10.7a-ii shows that when Ph-LFM is subjected to coherent interference, a ghost target appears. However, with

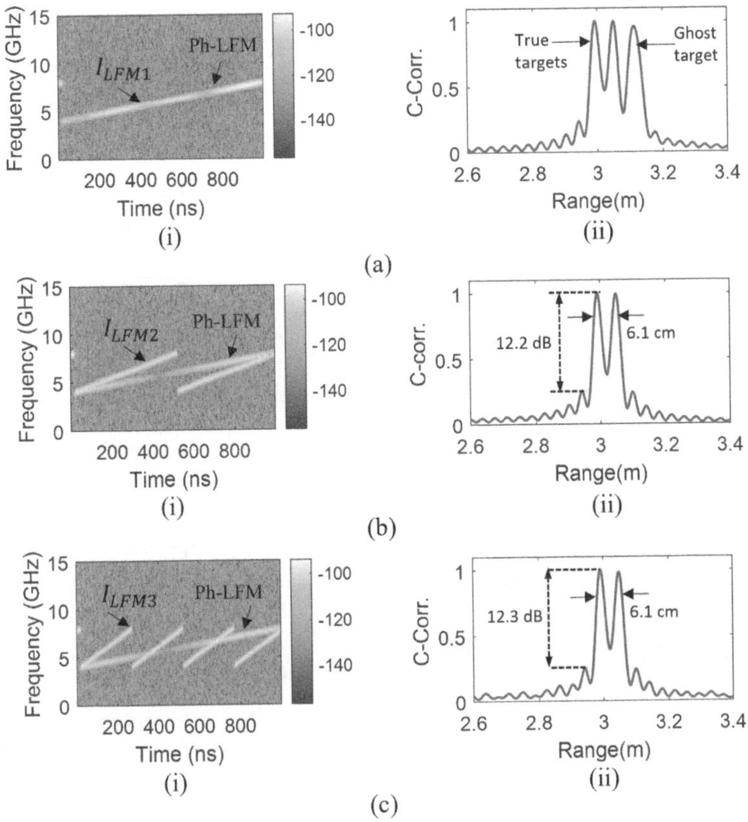

FIGURE 10.7 Illustrations of objects detection under interference signals with varying chirp rates for the Ph-LFM waveform, with chirp rates of (a) 4 GHz/μs, (b) 8 GHz/μs, and (c) 16 GHz/μs. (i) frequency-time representation and (ii) range profile results [50].

I_{LFM2} and I_{LFM3}, the ghost target is not observed. Despite this, the PSL value is reduced by 0.5 dB compared to the scenario without interference.

A similar analysis is performed on the Ph-O-MLFM signal, as shown in Figure 10.8. In the range profile presented in Figure 10.8a-ii, the ghost target signal no longer appears, unlike in Figure 10.7a-ii. For all IR signals depicted in Figure 10.8a–c, a PSL of over 11 dB and a −3 dB pulse width of 4 cm are achieved. The objects remain resolvable without significant degradation compared to the scenario without interference, as illustrated in Figure 10.6b. Only a 0.4-dB decline in the PSL is observed, indicating the Ph-O-MLFM signal's effectiveness in mitigating interference. The obtained

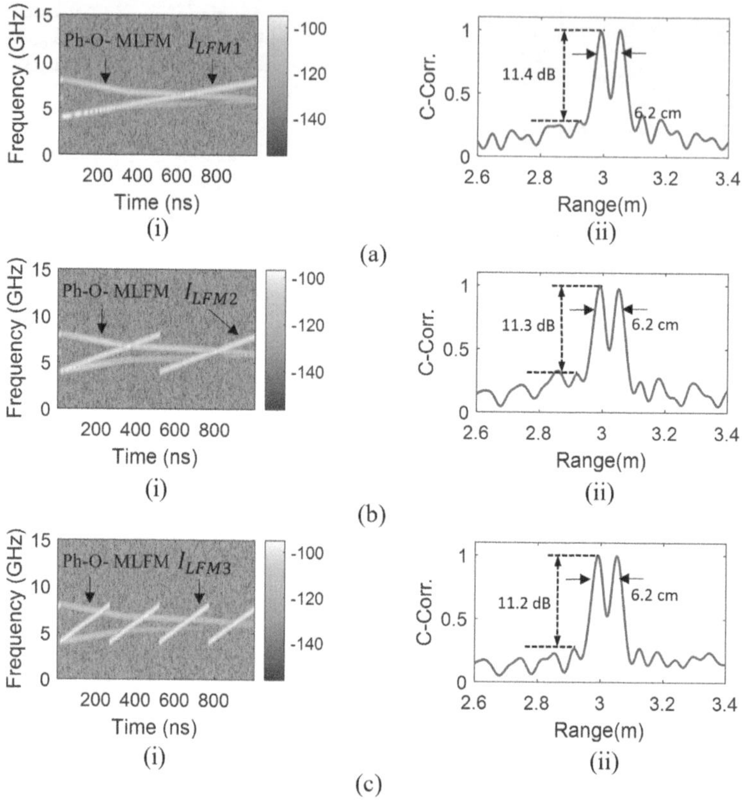

FIGURE 10.8 Illustrations of object detection under interference signals with varying chirp rates for the Ph-O-MLFM waveform, with chirp rates of (a) 4 GHz/μs, (b) 8 GHz/μs, and (c) 16 GHz/μs. (i) frequency-time representation and (ii) range profile results [50].

range resolution of 4 cm is close to the theoretical range resolution of 3.87 cm for a 4-GHz LFM signal.

It was observed that in the obtained range profile, the Ph-LFM waveform suffers with coherent interference. However, the Ph-O-MLFM waveform effectively mitigates these interferences, achieving a PSL of over 11 dB and a range resolution of 4 cm, which is close to the theoretical range resolution value of the 4 GHz LFM signal. These results show the Ph-O-MLFM waveform's suitability for use in multi-radar environments, offering excellent resolution and inherent resistance to interference.

10.3.2.2 Radar-Radar Interference Mitigation with Deep Learning

As in other fields, DL techniques have recently attracted considerable attention from the radar research community, particularly for IM. Among the various DL-IM techniques, convolutional neural network (CNN)-based methods [52–55] and recurrent neural network (RNN)-based methods [56–58] are commonly studied. In radar systems, IR signals distort the original echo signal's time domain sequence by adding undesired interference components, which need to be suppressed. RNN-based techniques effectively address these sequence-based temporal regression problems and provide promising solutions. In this chapter, we introduce a long short-term memory (LSTM)-based multi-layer recurrent neural network (LSTM-RNN) to address interference issues in a multi-radar environment [59].

In Refs. [38,60], the authors introduced a robust LSTM-based RNN (LSTM-RNN) to suppress interference signals from the received echo. The setup shown in Figure 10.9 was created to demonstrate IM using LSTM-RNN. A VR LFM transmission signal with a bandwidth of 4 GHz (4–8 GHz) was generated. The VR signal was employed to detect two objects located 3 m away from the radar sensor, with a separation of 4 cm between each object.

Table 10.1 presents the parameters for VR and IR signals. The VR signal is generated with a 1-μs time period and a 4-GHz bandwidth. This VR signal is used to detect two objects, and the received echo serves as the output label for the LSTM-RNN. The external electronically generated LFM interference signals were created with uniformly distributed random chirp rates (1–20 GHz/μs), bandwidths (1–6 GHz), and AWGN noise equivalent to SNR of (0–20 dB) as shown in Table 10.1. A total of 70 interference signals with varying parameters were generated. To prepare the training and test datasets, the LFM interference signals were mixed with the echo signals.

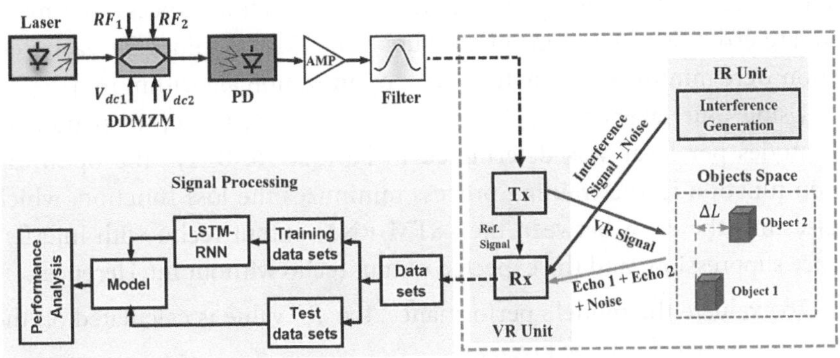

FIGURE 10.9 Block diagram for generating VR and IR radar signals and the use of LSTM-RNN for mitigating multi-radar interference [60].

TABLE 10.1 Dataset Preparation with VR and IR

IR and VR Signals Generation		
Parameters	**VR**	**IR**
No. of signals	1	70
Bandwidth	4 GHz	U{1–6} GHz
Chirp rates	4 GHz/μs	U{±1 to ±20} GHz/μs
Amplitude scaling	1	U{0.1–3}
Noise	-	AWGN equivalent to SNR of U {0 to +20} dB

TABLE 10.2 Optimum Hyperparameters

Hyperparameters	Values
No. of layers	4
Nodes per layer	(32,16,8,1)
Learning rate	0.004
Batch size	32
Epoch	25

For testing purposes, the test signals with one to eight interference signals were selected. For testing purpose, different interference signals from those used in the training stage were chosen randomly to ensure that the trained model performs properly with various interference signals with different parameters.

Before applying the data signals to the LSTM-RNN, they were normalized with zero mean and unit variance normalization. Optimal hyperparameter values were determined through hyperparameter tuning using the grid search method. Obtained optimal hyperparameters are presented in Table 10.2. During the hyperparameters optimization process, the mean square error (MSE) and mean absolute error (MAE) values for each iteration were minimized using the Adam optimization algorithm [61].

Using four layers and a learning rate of 0.004, the optimal number of nodes per layer was determined to be (32, 16, 8, 1). The optimization function in the training process minimizes the loss function, which measures the error between the LSTM-RNN output (echo with interference suppression) and the expected output (echo without interference).

To evaluate the model's performance, the R^2 value is calculated on the test signals which can be given as

$$R^2 = \left[1 - \frac{1}{N}\sum_{n=1}^{N}\left(y[n] - \tilde{y}[n]\right)^2\right] \tag{10.5}$$

where, N represents the sample length, $y[n]$ denotes the expected value, and $\tilde{y}[n]$ denotes the calculated value by LSTM-RNN. In radar systems, external interference usually causes two main issues in the range profile: (i) an increase in the noise floor due to non-coherent interference, and (ii) the appearance of ghost targets due to coherent interference [62–64]. SINR and EVM values are used as quantitative measures of radar detection performance. Therefore, to assess the similarity in the range profiles between echo signals without interference and interference-suppressed echo signals, EVM and SINR calculations were conducted. The EVM value can be calculated as [54]

$$\text{EVM} = \frac{1}{N}\sum_{n=1}^{N}\frac{\left|S[n] - \tilde{S}[n]\right|}{\left|S[n]\right|} \tag{10.6}$$

where, $|x|$ denotes the absolute value of the signal x, $\tilde{S}[n]$ represents the normalized range profile value obtained for interference-suppressed echo signal, while $S[n]$ denotes the value for the normalized range profile of the reference echo signal. Additionally, the SINR can be derived from EVM as: $\text{SINR}[\text{dB}] = 10 \times \text{Log}(1/\text{EVM})$. The SINR calculation accounts not only for the remaining interference but also for all distortions present in the interference-suppressed echo signal relative to the reference echo signal [54].

Figure 10.10 shows the spectrograms at various stages. Figure 10.10a-i and b-i show the frequency-time diagrams of the transmitted and received echo signals, respectively. Figure 10.10b-i illustrates the frequency-time diagrams with multiple coherent and non-coherent interferences affecting the echo signal. Similarly, Figure 10.10c-i depicts the effect of multiple coherent interferences on the echo signal. Figure 10.10b-ii and c-ii show the recovered echo signals after applying the LSTM-RNN to the cases in Figure 10.10b-i and c-i, respectively. As shown in Figure 10.10b-ii and c-ii, the LSTM-RNN effectively suppresses multiple interferences.

Figure 10.11 presents the range profile results for different test scenarios. Coherent and non-coherent interference signals are added to the victim

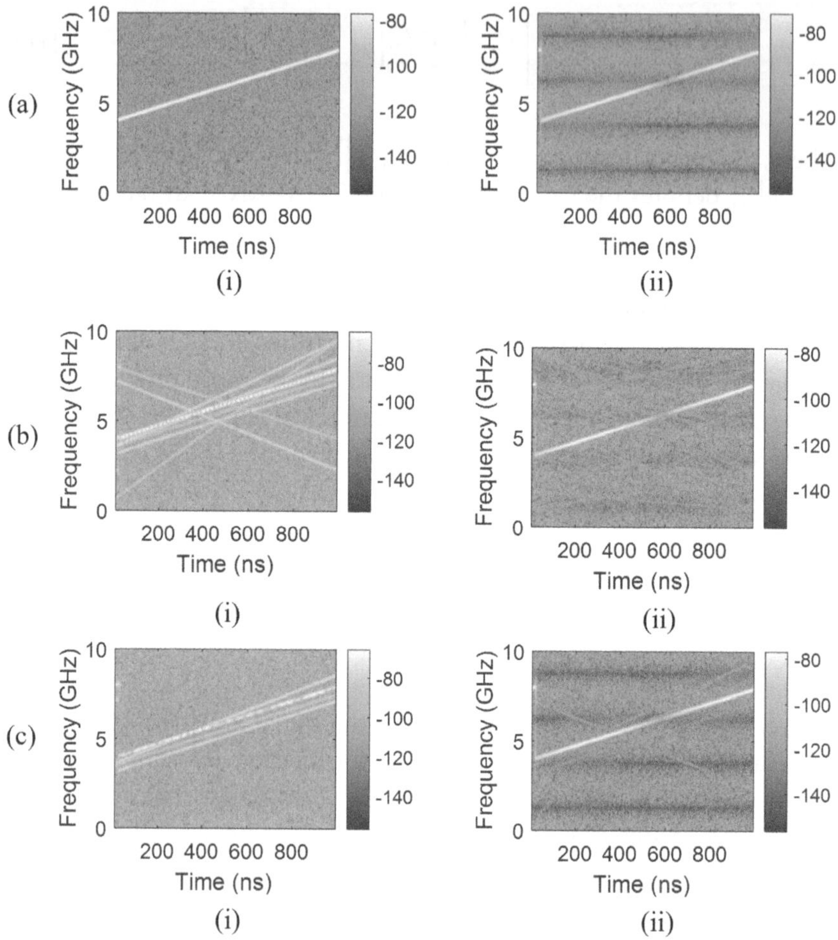

FIGURE 10.10 Frequency-time diagrams. (a) VR signal; (i) Tx signal and (ii) echo signal. Spectrograms of echo signal before applying LSTM-RNN; (b-i) with multiple mixed interferences and (c-i) with multiple coherent interferences. (b-ii) and (c-ii) spectrograms after interferences mitigation with LSTM-RNN.

echo signals for testing purposes. Figure 10.11a corresponds to the case of both coherent and non-coherent interferences, as shown in Figure 10.10b, while Figure 10.11b corresponds to the case of coherent interferences acting upon the echo signal, as shown in Figure 10.10c. In both figures, the blue color represents the range profile results after mitigating interference with the LSTM-RNN, and the red color represents the range profile results with interferences. As observed, the interference significantly

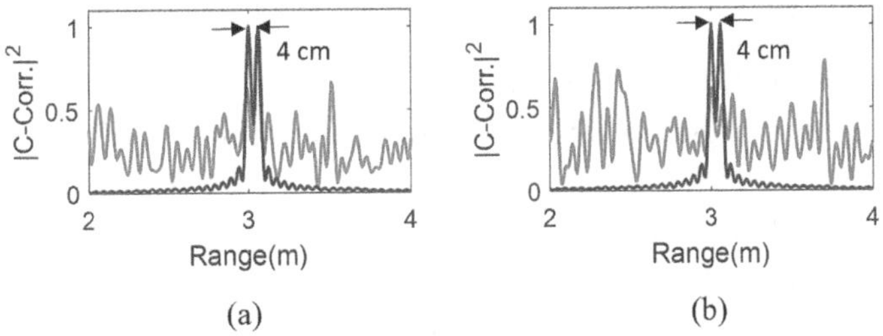

FIGURE 10.11 Range profiles: (a) with eight mixed interferences and (b) with eight coherent interferences. Dark grey color denotes range profile after mitigating the interference with LSTM-RNN and light grey waveform denotes with interferences.

TABLE 10.3 Model Evaluation

No. of Interferences	R2	MSE	MAE	EVM	SINR [dB]
1	92.8%	0.09	0.25	0.066	11.9
4	92.7%	0.091	0.26	0.068	11.81
8	92.4%	0.094	0.28	0.073	11.64

increases the noise floor. However, the LSTM-RNN effectively suppresses the interference-induced noise. The quantitative measure of effectiveness of the LSTM-RNN can be evaluated using R^2, SINR, and EVM values.

Table 10.3 summarizes the average results (based on 15 trials) for test signals with 1, 4, and 8 interference signals. The same previously trained model was used to evaluate each test signal. During testing, the loss metrics (MSE / MAE) were calculated using the trained model. The results in Table 10.3 show that the designed model can recover the interference-suppressed echo signal from the received echo signal with multiple interference signals acting on the VR, achieving a mean R^2 value greater than 92%. For the given signal scenario, the optimal hyperparameters resulted in an LSTM-RNN model with four layers, sufficient to achieve an R^2 value of over 92% for all test data, demonstrating strong performance for regression problems. Thus, the obtained hyperparameters are optimal for the dataset used. Additionally, EVM values below 0.075 and SINR values above 11 dB were achieved for test signals with 1, 4, and 8 interferences, which are acceptable for practical scenarios [54].

REFERENCES

[1] J. Bruder, J. Carlo, J. Gurney, and J. Gorman, "IEEE standard for letter designations for radar-frequency bands," *IEEE Aerospace & Electronic Systems Society*, vol. 6, pp. 1–3, 2003.

[2] S. Z. Gurbuz, H. D. Griffiths, A. Charlish, M. Rangaswamy, M. S. Greco, and K. Bell, "An overview of cognitive radar: Past, present, and future," *IEEE Aerospace and Electronic Systems Magazine,* vol. 34, pp. 6–18, 2019.

[3] H. Griffiths, L. Cohen, S. Watts, E. Mokole, C. Baker, M. Wicks, et al., "Radar spectrum engineering and management: Technical and regulatory issues," *Proceedings of the IEEE,* vol. 103, pp. 85–102, 2014.

[4] F. Engels, P. Heidenreich, M. Wintermantel, L. Stäcker, M. Al Kadi, and A. M. Zoubir, "Automotive radar signal processing: Research directions and practical challenges," *IEEE Journal of Selected Topics in Signal Processing,* vol. 15, pp. 865–878, 2021.

[5] S. Pisa, E. Pittella, and E. Piuzzi, "A survey of radar systems for medical applications," *IEEE Aerospace and Electronic Systems Magazine,* vol. 31, pp. 64–81, 2016.

[6] S. C. Steele-Dunne, H. McNairn, A. Monsivais-Huertero, J. Judge, P.-W. Liu, and K. Papathanassiou, "Radar remote sensing of agricultural canopies: A review," *IEEE Journal of Selected Topics in Applied Earth Observations and Remote Sensing,* vol. 10, pp. 2249–2273, 2017.

[7] A. D. Brown, "Radar challenges, current solutions, and future advancements for the counter unmanned aerial systems mission," *IEEE Aerospace and Electronic Systems Magazine,* vol. 38, pp. 34–50, 2023.

[8] S. Chaudhary, L. Wuttisittikulkij, M. Saadi, A. Sharma, S. Al Otaibi, J. Nebhen, et al., "Coherent detection-based photonic radar for autonomous vehicles under diverse weather conditions," *PLoS One,* vol. 16, p. e0259438, 2021.

[9] A. Sharma, S. Chaudhary, J. Malhotra, M. Saadi, S. Al Otaibi, J. Nebhen, et al., "A cost-effective photonic radar under adverse weather conditions for autonomous vehicles by incorporating a frequency-modulated direct detection scheme," *Frontiers in Physics,* vol. 9, p. 467, 2021.

[10] A. Sharma and J. Malhotra, "Simulative investigation of FMCW based optical photonic radar and its different configurations," *Optical and Quantum Electronics,* vol. 54, p. 233, 2021.

[11] P. Sood, A. Sharma, and Chandni, "Analysis of FSO system and its challenges–A review," *International Journal of Computer Applications,* vol. 179, pp. 42–45, 2018.

[12] Z. Zhou, H. Zhang, C. Lin, and A. Sharma, "Performance analysis of duobinary and CSRZ modulation based polarization interleaving for high-speed WDM-FSO transmission system," *Journal of Optical Communications,* vol. 43, pp. 147–152, 2018.

[13] K. Shakthi Murugan, A. Sharma, and J. Malhotra, "Performance analysis of 80 Gbps Ro-FSO system by incorporating hybrid WDM-MDM scheme," *Optical and Quantum Electronics,* vol. 52, pp. 1–12, 2020.

[14] A. Sharma and V. Kumar, "High speed and long reach DPSK-OFDM-Is-OWC under impact of space turbulences," *International Journal of Computer Applications,* vol. 975, p. 8887, 2018.

[15] A. Sharma and R. Kapoor, "Study of various challenges in is OWC: A review," *International Journal for Research in Applied Science and Engineering Technology (IJRASET),* vol. 5, pp. 802–807, 2017.

[16] S. Chaudhary, D. Thakur, and A. Sharma, "10 Gbps-60 GHz RoF transmission system for 5 G applications," *Journal of Optical Communications,* vol. 40, pp. 281–284, 2017.

[17] S. Chaudhary, P. Chauhan, and A. Sharma, "High speed 4 × 2.5 Gbps-5 GHz AMI-WDM-RoF transmission system for WLANs," *Journal of Optical Communications,* vol. 40, pp. 285–288, 2019.

[18] S. Chaudhary, D. Thakur, and A. Sharma, "10 Gbps-60 GHz RoF transmission system for 5 G applications," *Journal of Optical Communications,* vol. 40, pp. 281–284, 2019.

[19] A. Sharma, V. Mishra, K. Singh, and J. Malhotra, "Hybrid RoF-RoFSO system for broadband services by incorporating polarization division multiplexing scheme," *Journal of Optical Communications,* 2023. https://doi.org/10.1515/joc-2023-0309

[20] S. S. S. Panda, T. Panigrahi, S. R. Parne, S. L. Sabat, and L. R. Cenkeramaddi, "Recent advances and future directions of microwave photonic radars: A review," *IEEE Sensors Journal,* vol. 21, pp. 21144–21158, 2021.

[21] P. Ghelfi, F. Laghezza, F. Scotti, G. Serafino, S. Pinna, D. Onori, et al., "Photonics in radar systems: RF integration for state-of-the-art functionality," *IEEE Microwave Magazine,* vol. 16, pp. 74–83, 2015.

[22] S. Pan and Y. Zhang, "Microwave photonic radars," *Journal of Lightwave Technology,* vol. 38, pp. 5450–5484, 2020.

[23] S. Chaudhary, A. Sharma, S. Khichar, X. Tang, X. Wei, and L. Wuttisittikulkij, "High resolution-based coherent photonic radar sensor for multiple target detections," *Journal of Sensor and Actuator Networks,* vol. 11, p. 49, 2022.

[24] A. Sharma, S. Chaudhary, J. Malhotra, S. Khichar, and L. Wuttisittikulkij, "Photonic sensor for multiple targets detection under adverse weather conditions in autonomous vehicles," *Journal of Sensor and Actuator Networks,* vol. 11, p. 60, 2022.

[25] A. Sharma, S. Chaudhary, J. Malhotra, A. Parnianifard, and L. Wuttisittikulkij, "Measurement of target range and Doppler shift by incorporating PDM-enabled FMCW-based photonic radar," *Optik,* vol. 262, p. 169191, 2022.

[26] A. Sharma, S. Chaudhary, J. Malhotra, M. Saadi, S. Al Otaibi, and L. Wuttisittikulkij, "Speed-direction sensing under multiple vehicles scenario using photonic radars," *CMC-Computers Materials & Continua,* vol. 73, pp. 5399–5410, 2022.

[27] Z. Zhang, Y. Liu, and B. J. Eggleton, "Photonic generation of 30 GHz bandwidth stepped-frequency signals for radar applications," *Journal of Lightwave Technology,* vol. 40, pp. 4521–4527, 2022.

[28] A. Sharma and J. Malhotra, "Simulative investigation of FMCW based optical photonic radar and its different configurations," *Optical and Quantum Electronics,* vol. 54, p. 233, 2022.

[29] A. Sharma and J. Malhotra, "Performance enhancement of photonic radar sensor for detecting multiple targets by incorporating mode division multiplexing," *Optical and Quantum Electronics,* vol. 54, p. 410, 2022.

[30] A. Sharma and J. Malhotra, "Evaluating the effects of material reflectivity and atmospheric attenuation on photonic radar performance in free space optical channels," *Journal of Optical Communications,* vol. 6, p. 1–6, 2023.

[31] A. Sharma, J. Malhotra, S. Khichar, A. Parnianifard, and S. Chaudhary, "Highly efficient frequency modulated continuous

wave based photonic radar by incorporating electronic equalization scheme," *Optical and Quantum Electronics*, vol. 55, p. 797, 2023.

[32] E. A. Kittlaus, D. Eliyahu, S. Ganji, S. Williams, A. B. Matsko, K. B. Cooper, et al., "A low-noise photonic heterodyne synthesizer and its application to millimeter-wave radar," *Nature Communications*, vol. 12, p. 4397, 2021.

[33] Q. Guo, F. Zhang, P. Zhou, and S. Pan, "Dual-band LFM signal generation by optical frequency quadrupling and polarization multiplexing," *IEEE Photonics Technology Letters*, vol. 29, pp. 1320–1323, 2017.

[34] F. Zhang, Q. Guo, Z. Wang, P. Zhou, G. Zhang, J. Sun, et al., "Photonics-based broadband radar for high-resolution and real-time inverse synthetic aperture imaging," *Optics Express*, vol. 25, pp. 16274–16281, 2017.

[35] S.-C. Chan, S.-K. Hwang, and J.-M. Liu, "Period-one oscillation for photonic microwave transmission using an optically injected semiconductor laser," *Optics Express*, vol. 15, pp. 14921–14935, 2007.

[36] X.-Q. Qi and J.-M. Liu, "Dynamics scenarios of dual-beam optically injected semiconductor lasers," *IEEE Journal of Quantum Electronics*, vol. 47, pp. 762–769, 2011.

[37] B. Nakarmi, B. Y. Song, H. N. Parajuli, I. A. Ukaegbu, W. Xiangchun, and S. Pan, "Multi-chirp piecewise linearly frequency modulated microwave generation using a semiconductor laser," *IEEE Photonics Technology Letters*, vol. 35, pp. 821–824, 2023.

[38] H. N. Parajuli, G. Bakhtiyarov, B. Nakarmi, and I. A. Ukaegbu, "Interference mitigation in multi-radar environment using LSTM-based recurrent neural network," in *International Conference on Intelligent Human Computer Interaction*, Ahmedabad, India, 2023, pp. 151–161.

[39] W.-Q. Wang and H. Shao, "Radar-to-radar interference suppression for distributed radar sensor networks," *Remote Sensing*, vol. 6, pp. 740–755, 2014.

[40] G. M. Brooker, "Mutual interference of millimeter-wave radar systems," *IEEE Transactions on Electromagnetic Compatibility*, vol. 49, pp. 170–181, 2007.

[41] J. Bechter and C. Waldschmidt, "Automotive radar interference mitigation by reconstruction and cancellation of interference component," in *2015 IEEE MTT-S International Conference on*

Microwaves for Intelligent Mobility (ICMIM), Heidelberg, Germany, 27–29 April, 2015, pp. 1–4.

[42] S. Chaudhary, A. Sharma, S. Khichar, Y. Meng, and J. Malhotra, "Enhancing autonomous vehicle navigation using SVM-based multi-target detection with photonic radar in complex traffic scenarios," *Scientific Reports,* vol. 14, p. 17339, 2024.

[43] S. Chaudhary, A. Sharma, M. A. Naeem, and Y. Meng, "Target detection in challenging environments: Photonic radar with a hybrid multiplexing scheme for 5G autonomous vehicles," *Sustainability,* vol. 16, p. 991, 2024.

[44] S. Chaudhary, A. Sharma, K. Singh, S. Khichar, and J. Malhotra, "Highly efficient photonic radar by incorporating MDM-WDM and machine learning classifiers under adverse weather conditions," *Plos One,* vol. 19, p. e0300653, 2024.

[45] Z. Xu and Q. Shi, "Interference mitigation for automotive radar using orthogonal noise waveforms," *IEEE Geoscience and Remote Sensing Letters,* vol. 15, pp. 137–141, 2017.

[46] F. Uysal, "Phase-coded FMCW automotive radar: System design and interference mitigation," *IEEE Transactions on Vehicular Technology,* vol. 69, pp. 270–281, 2019.

[47] Y.-S. Son, H.-K. Sung, and S. W. Heo, "Automotive frequency modulated continuous wave radar interference reduction using per-vehicle chirp sequences," *Sensors,* vol. 18, p. 2831, 2018.

[48] C. Ma, Y. Yang, F. Cao, X. Wang, X. Liu, C. Meng, et al., "High-resolution microwave photonic radar with sparse stepped frequency chirp signals," *IEEE Transactions on Geoscience and Remote Sensing,* vol. 60, pp. 1–10, 2022.

[49] F.-Y. Lin and J.-M. Liu, "Chaotic radar using nonlinear laser dynamics," *IEEE Journal of Quantum Electronics,* vol. 40, pp. 815–820, 2004.

[50] H. N. Parajuli, B. Nakarmi, A. Ashimbayeva, and I. A. Ukaegbu, "Interference-resistant photonically generated orthogonal pair multi-chirp-LFM waveform," in *2023 International Topical Meeting on Microwave Photonics (MWP), Nanjing, China, 15–18 October,* 2023, pp. 1–4.

[51] G. Galati and G. Pavan, "Orthogonal and complementary radar signals for multichannel applications," in *2011 8th European Radar Conference, Manchester, UK, 12–14 October,* 2011, pp. 178–181.

[52] T. Oyedare, V. K. Shah, D. J. Jakubisin, and J. H. Reed, "Interference suppression using deep learning: Current approaches and open challenges," *IEEE Access,* vol. 10, pp. 66238–66266, 2022.

[53] J. Rock, W. Roth, M. Toth, P. Meissner, and F. Pernkopf, "Resource-efficient deep neural networks for automotive radar interference mitigation," *IEEE Journal of Selected Topics in Signal Processing,* vol. 15, pp. 927–940, 2021.

[54] J. Wang, R. Li, X. Zhang, and Y. He, "Interference mitigation for automotive FMCW radar based on contrastive learning with dilated convolution," *IEEE Transactions on Intelligent Transportation Systems,* vol. 25, pp. 545–558, 2023.

[55] J. Fuchs, A. Dubey, M. Lübke, R. Weigel, and F. Lurz, "Automotive radar interference mitigation using a convolutional autoencoder," in *2020 IEEE International Radar Conference (RADAR), Washington, DC, 28–30 April,* 2020, pp. 315–320.

[56] M. Rameez, S. Javadi, M. Dahl, and M. I. Pettersson, "Signal reconstruction using Bi-LSTM for automotive radar interference mitigation," in *2021 18th European Radar Conference (EuRAD), London, 5–7 April,* 2022, pp. 74–77.

[57] J. Hille, D. Auge, C. Grassmann, and A. Knoll, "FMCW radar2radar interference detection with a recurrent neural network," in *2022 IEEE Radar Conference (RadarConf22), New York City, NY, 21–25 March,* 2022, pp. 1–6.

[58] R. Koizumi, X. Wang, M. Umehira, S. Takeda, and R. Sun, "RNN-based interference suppression method for CS radar: Simulation and experimental evaluations," in *2023 International Conference on Artificial Intelligence in Information and Communication (ICAIIC), Bali, Indonesia, 20–23 February,* 2023, pp. 242–247.

[59] S. Hochreiter, "*Long Short-term Memory,*" *Neural Computation MIT-Press,* 1997.

[60] H. Parajuli, A. Ashimbayeva, U. Nakarmi, I. Ukaegbu, B. Gaudel, S. Pan, et al., "Multi-radar interference mitigation in photonics-based radar with sliding window LSTM recurrent neural network," *Journal of Lightwave Technology,* vol. 42, pp. 7567–7576, 2024.

[61] D. Kinga and J. B. Adam, "A method for stochastic optimization," in *International Conference on Learning Representations (ICLR),* San Juan, Puerto Rico, 2015, p. 6.

[62] M. Goppelt, H.-L. Blöcher, and W. Menzel, "Analytical investigation of mutual interference between automotive FMCW radar sensors," in *2011 German Microwave Conference, Darmstadt, Germany, 14–15 March*, 2011, pp. 1–4.

[63] G. Kim, J. Mun, and J. Lee, "A peer-to-peer interference analysis for automotive chirp sequence radars," *IEEE Transactions on Vehicular Technology*, vol. 67, pp. 8110–8117, 2018.

[64] M. Toth, P. Meissner, A. Melzer, and K. Witrisal, "Analytical investigation of non-coherent mutual FMCW radar interference," in *2018 15th European Radar Conference (EuRAD), Madrid, Spain, 26–28 September*, 2018, pp. 71–74.

Index